ELECTRICAL ENGINEERING: CONCEPTS AND APPLICATIONS

Edited by
Rongie N. Bolanos

Delve Publishing

Electrical Engineering: Concepts and Applications

Edited by Rongie N. Bolanos

Edition 2018

Delve Publishing

Published by
Delve Publishing LLC
708 3rd Avenue, 6th Floor
New York
NY 10801
United States of America
www.delvepublishing.com

ISBN: 978-1-68095-684-9 (Hardback)

Printed in United States of America

Library of Congress Control Number: 2016920457

About the Editor

Rongie N. Bolanos

Rongie N. Bolanos finished Bachelor of Science in Electronics and Communication Engineering at the Rizal Technological University and is currently working as drafter.

Table of Contents

2 DEVELOPMENTS IN ELECTRICAL ENGINEERING

3 ELECTROMAGNETISM

4 POWER ENGINEERING BASICS

5 ELECTRIC MOTOR

6 TRANSFORMER

7 SWITCHGEAR AND PROTECTION

8 ELECTRICAL POWER TRANSMISSION

9 INSTRUMENTATION

List of Abbreviations

AC	alternating current
ECMs	electronically commutated motors
ERCOT	Electric Reliability Council of Texas
EMI	electromagnetic interference
EMF	electromotive force
CO2	Carbon Dioxide
GIS	gas-insulated switchgear
DC	direct current
HVDC	high-voltage direct current
IEEE	Institute of Electrical and Electronics Engineers
IET	Institution of Engineering and Technology
LFH	low-frequency heating
ME	Metal-enclosed
MEMS	Microelectromechanical systems
SF	sulfur hexafluoride
USS	unitized substation
PD	partial discharge
PM	permanent magnet
PLCs	programmable logic controllers
VPD	vapor-phase drying
VPE	vacuum pressure encapsulation
VPI	vacuum pressure impregnation
VA	volt-ampere

Preface

Electrical engineering is concerned with research, design, development, manufacture, installation, operation, maintenance and management of equipment, plant and systems within the electrical, electronic, communication and computer systems areas. These activities can apply to electricity generation, transmission, distribution, electrical installations in buildings and on industrial sites, electrical equipment manufacture, instrumentation and control systems applications in industry, communications networks, electronic plant and equipment, and also the integration and control of computer systems. Since its early beginnings, the field of electrical engineering has grown and branched out into a number of specialized categories, including power generation and transmission systems, motors, batteries and control systems. Electrical engineering also includes electronics, which has itself branched into an even greater number of subcategories, such as radio frequency (RF) systems, telecommunications, remote sensing, signal processing, digital circuits, instrumentation, audio, video and optoelectronics. The field of electronics was born with the invention of the thermionic valve diode vacuum tube in 1904 by John Ambrose Fleming. The vacuum tube basically acts as a current amplifier by outputting a multiple of its input current. It was the foundation of all electronics, including radios, television and radar, until the mid-20th century. It was largely supplanted by the transistor, which was developed in 1947 at AT&T's Bell Laboratories by William Shockley, John Bardeen and Walter Brattain, for which they received the 1956 Nobel Prize in physics. Electrical engineers are increasingly relying on computer-aided design (CAD) systems to create schematics and lay out circuits. They also use computers to simulate how electrical devices and systems will function. Computer simulations can be used to model a national power grid or a microprocessor; therefore, proficiency with computers is essential for electrical engineers. In addition to speeding up the process of drafting schematics, printed circuit board (PCB) layouts and blueprints for electrical and electronic devices, CAD systems allow for quick and easy modifications of designs and rapid prototyping using CNC machines.

Content Coverage

Chapters One and *Two* present an overview of electricity and developments in electrical engineering, respectively. Electricity gives a wide variety of well-known effects, such as lightning, static electricity, electromagnetic induction and electric current.

Chapters Three and *Four* provide coverage of electromagnetism which deals with electricity and magnetism and the interaction between them, and power systems engineering that deals with the generation, transmission, distribution and utilization of electric power and the electrical devices connected to such systems including generators, motors and transformers, separately.

Chapters Five and *Six* are electric motor and transformer, correspondingly. Electric motors are used to produce linear or rotary force (torque), and should be distinguished from devices such as magnetic solenoids and loudspeakers that convert electricity into motion but do not generate usable mechanical powers, which are respectively referred to as actuators and transducers. A transformer is an electrical device that transfers electrical energy between two or more circuits through electromagnetic induction.

Chapters Seven and *Eight* present about switchgear and electrical power transmission, separately. Switchgear is used both to de-energize equipment to allow work to be done and to clear faults downstream.

Chapter Nine covers measuring instruments used for indicating, measuring and recording physical quantities.

CHAPTER 1

Electrical Electricity Fundamentals

Chapter Contents

Electrical results from the movement of an electrical charge, and is commonly referred to as simply "electricity." Ultimately, it has its origin in the electromagnetic force: one of the four fundamental forces of nature and the one that is responsible for the behavior of electrically charged objects. Electrical energy is the result of the interaction of subatomic particles with this force. Electricity manifests itself in natural phenomena such as lightning and is essential to life at a fundamental level. The ability of humans to generate, transmit and store electricity is crucial to modern industry, technology and, in most countries, domestic life.

INTRODUCTION

Electricity is the set of physical phenomena associated with the presence and flow of electric charge. Electricity gives a wide variety of well-known effects, such as lightning, static electricity, electromagnetic induction and electric current. In addition, electricity permits the creation and reception of electromagnetic radiation such as radio waves.

In electricity, charges produce electromagnetic fields which act on other charges. Electricity occurs due to several types of physics:

- Electric charge: a property of some subatomic particles, which determines their electromagnetic interactions. Electrically charged matter is influenced by, and produces, electromagnetic fields, electric charges can be positive or negative.
- Electric field (see electrostatics): charges are surrounded by an electric field. The electric field produces a force on other charges. Changes in the electric field travel at the speed of light.
- Electric potential: the capacity of an electric field to do work on an electric charge, typically measured in volts.
- Electric current: a movement or flow of electrically charged particles, typically measured in amperes.
- Electromagnets: moving charges produce a magnetic field. Electric currents generate magnetic fields, and changing magnetic fields generate electric currents.

In electrical engineering, electricity is used for:

- Electric power where electric current is used to energise equipment;
- Electronics which deals with electrical circuits that involve active electrical components such as vacuum Tubes, transistors, diodes and integrated circuits, and associated passive interconnection technologies.

Electrical phenomena have been studied since antiquity, though progress in theoretical understanding remained slow until the seventeenth and eighteenth centuries. Even then, practical applications for electricity were few, and it would not be until the late nineteenth century that engineers were able to put it to industrial and residential use. The rapid expansion in electrical technology at this time transformed industry and society. Electricity's extraordinary versatility means it can be put to an almost limitless set of applications which include transport, heating, lighting, communications, and computation. Electrical power is now the backbone of modern industrial society.

Lightning is one of the most dramatic effects of electricity.

ELECTRIC CHARGE

Electric charge is the physical property of matter that causes it to experience a force when placed in an electromagnetic field. There are two types of electric charges: positive and negative. Like charges repel and unlike attract. An object is negatively charged if it has an excess of electrons, and is otherwise positively charged or uncharged. The SI derived unit of electric charge is the coulomb (C). In electrical engineering, it is also common to use the ampere-hour (Ah), and, in chemistry, it is common to use the elementary charge (*e*) as a unit. The symbol *Q* often denotes charge. Early knowledge of how charged substances interact is now called classical electrodynamics, and is still accurate for problems that don`t require consideration of quantum effects.

The *electric charge* is a fundamental conserved property of some subatomic particles, which determines their electromagnetic interaction. Electrically charged matter is influenced by, and produces, electromagnetic fields. The interaction between a moving charge and an electromagnetic field is the source of the electromagnetic force, which is one of the four fundamental forces .

Twentieth-century experiments demonstrated that electric charge is *quantized*; that is, it comes in integer multiples of individual small units called the elementary charge, *e*, approximately equal to 1.602×10^{-19} coulombs (except for particles called quarks, which have charges that are integer multiples of $\frac{1}{3}e$). The proton has a charge of +*e*, and the electron has a charge of *e*. The study of charged particles, and how their interactions are mediated by photons, is called quantum

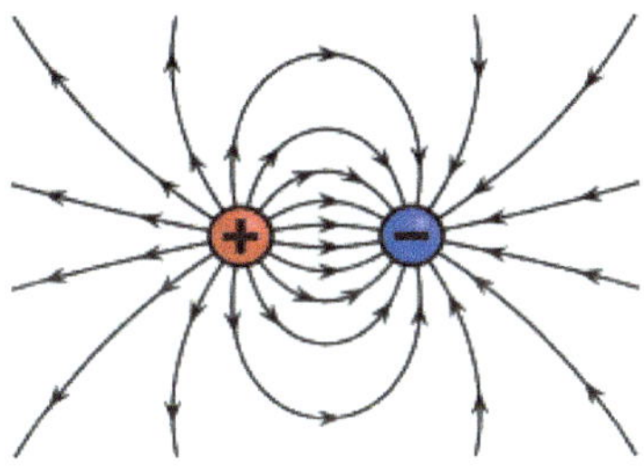

Electric field of a positive and a negative point charge.

Overview

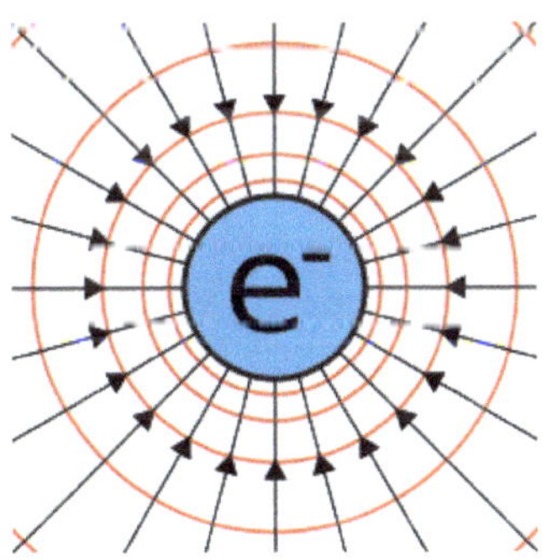

Diagram showing field lines and equipotentials around an electron, a negatively charged particle. In an electrically neutral atom, the number of electrons is equal to the number of protons (which are positively charged), resulting in a net zero overall charge

Charge is the fundamental property of forms of matter that exhibit electrostatic attraction or repulsion in the presence of other matter. Electric charge is a characteristic property of many subatomic particles. The charges of free-standing particles are integer multiples of the elementary charge *e*; we say that electric charge is *quantized*. Michael Faraday, in his electrolysis experiments, was the first to note the discrete nature of electric charge. Robert Millikan`s oil drop experiment demonstrated this fact directly, and measured the elementary charge.

By convention, the charge of an electron is −1, while that of a proton is +1. Charged particles whose charges have the same sign repel one another, and particles whose charges have different signs attract. Coulomb`s law quantifies the electrostatic force between two particles by asserting that the force is proportional to the product of their charges, and inversely proportional to the square of the distance between them.

The charge of an antiparticle equals that of the corresponding particle, but with opposite sign. Quarks have fractional charges of either −1/3 or +2/3, but free-standing quarks have never been observed (the theoretical reason for this fact is asymptotic freedom).

The electric charge of a macroscopic object is the sum of the electric charges of the particles that make it up. This charge is often small, because matter is made of atoms, and atoms typically have equal numbers of protons and electrons, in which case their charges cancel out, yielding a net charge of zero, thus making the atom neutral.

An *ion* is an atom (or group of atoms) that has lost one or more electrons, giving it a net positive charge (cation), or that has gained one or more electrons, giving it a net negative charge (anion). *Monatomic ions* are formed from single atoms, while *polyatomic ions* are formed from two or more atoms that have been bonded together, in each case yielding an ion with a positive or negative net charge.

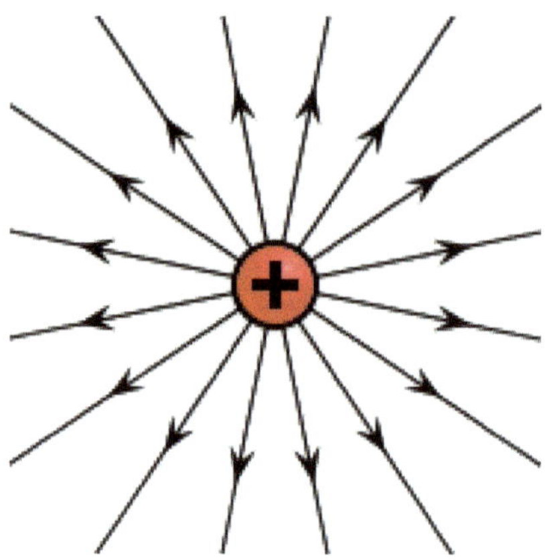

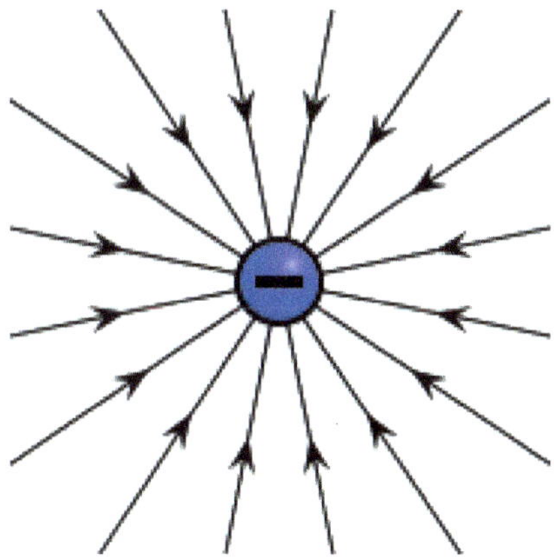

Electric field induced by a positive electric charge (left) and a field induced by a negative electric charge (right).

During formation of macroscopic objects, constituent atoms and ions usually combine to form structures composed of neutral *ionic compounds* electrically bound to neutral atoms. Thus macroscopic objects tend toward being neutral overall, but macroscopic objects are rarely perfectly net neutral.

Sometimes macroscopic objects contain ions distributed throughout the material, rigidly bound in place, giving an overall net positive or negative charge to the object. Also, macroscopic objects made of conductive elements, can more or less easily (depending on the element) take on or give off electrons, and then maintain a net negative or positive charge indefinitely. When the net electric charge of an object is non-zero and motionless, the phenomenon is known as static electricity. This can easily be produced by rubbing two dissimilar materials together, such as rubbing amber with fur or glass with silk. In this way non-conductive materials can be charged to a significant degree, either positively or negatively. Charge taken from one material is moved to the other material, leaving an opposite charge of the same magnitude behind. The law of *conservation of charge* always applies, giving the object from which a negative charge is taken a positive charge of the same magnitude, and vice versa.

Even when an object's net charge is zero, charge can be distributed non-uniformly in the object (e.g., due to an external electromagnetic field, or bound polar molecules). In such cases the object is said to be polarized. The charge due to polarization is known as bound charge, while charge on an object produced by electrons gained or lost from outside the object is called *free charge*. The motion of electrons in conductive metals in a specific direction is known as electric current.

Units

The SI unit of quantity of electric charge is the coulomb, which is equivalent to about 6.242×10^{18} *e* (*e* is the charge of a proton). Hence, the charge of an electron is approximately -1.602×10^{-19} C. The coulomb is defined as the quantity of charge that has passed through the cross section of an electrical conductor carrying one ampere within one second. The symbol Q is often used to denote a quantity of electricity or charge. The quantity of electric charge can be directly measured with an electrometer, or indirectly measured with a ballistic galvanometer.

After finding the quantized character of charge, in 1891 George Stoney proposed the unit 'electron' for this fundamental unit of electrical charge. This was before the discovery of the particle by J.J. Thomson in 1897. The unit is today treated as nameless, referred to as "elementary charge", "fundamental unit of charge", or simply as "e". A measure of charge should be a multiple of the elementary charge e, even if at large scales charge seems to behave as a real quantity. In some contexts it is meaningful to speak of fractions of a charge; for example in the charging of a capacitor, or in the fractional quantum Hall effect.

In systems of units other than SI such as cgs, electric charge is expressed as combination of only three fundamental quantities (length, mass, and time), and not four, as in SI, where electric charge is a combination of length, mass, time, and electric current.

Static Electricity and Electric Current

Static electricity and electric current are two separate phenomena. They both involve electric charge, and may occur simultaneously in the same object. Static electricity refers to the electric charge of an object and the related electrostatic discharge when two objects are brought together that are not at equilibrium. An electrostatic discharge creates a change in the charge of each of the two objects. In contrast, electric current is the flow of electric charge through an object, which produces no net loss or gain of electric charge.

Electrification by Friction

When a piece of glass and a piece of resin—neither of which exhibit any electrical properties—are rubbed together and left with the rubbed surfaces in contact, they still exhibit no electrical properties. When separated, they attract each other.

A second piece of glass rubbed with a second piece of resin, then separated and suspended near the former pieces of glass and resin causes these phenomena:

- The two pieces of glass repel each other.
- Each piece of glass attracts each piece of resin.
- The two pieces of resin repel each other.

This attraction and repulsion is an *electrical phenomena*, and the bodies that exhibit them are said to be *electrified*, or *electrically charged*. Bodies may be electrified in many other ways, as well as by friction. The electrical properties of the two pieces of glass are similar to each other but opposite to those of the two pieces of resin: The glass attracts what the resin repels and repels what the resin attracts.

If a body electrified in any manner whatsoever behaves as the glass does, that is, if it repels the glass and attracts the resin, the body is said to be *vitreously* electrified, and if it attracts the glass and repels the resin it is said to be *resinously* electrified. All electrified bodies are either vitreously or resinously electrified.

An established convention in the scientific community defines vitreous electrification as positive, and resinous electrification as negative. The exactly opposite properties of the two kinds of electrification justify our indicating them by opposite signs, but the application of the positive sign to one rather than to the other kind must be considered as a matter of arbitrary convention—just as it is a matter of convention in mathematical diagram to reckon positive distances towards the right hand.

No force, either of attraction or of repulsion, can be observed between an electrified body and a body not electrified.

Actually, all bodies are electrified, but may appear not electrified because of the relatively similar charge of neighboring objects in the environment. An object further electrified + or – creates an equivalent or opposite charge by default in neighboring objects, until those charges can equalize. The effects of attraction can be observed in high-voltage experiments, while lower voltage effects are merely weaker and therefore less obvious. The attraction and repulsion forces are codified by Coulomb`s law (attraction falls off at the square of the distance, which has a corollary for acceleration in a gravitational field, suggesting that gravitation may be merely electrostatic phenomenon between relatively weak charges in terms of scale).

It is now known that the Franklin-Watson model was fundamentally correct. There is only one kind of electrical charge, and only one variable is required to keep track of the amount of charge. On the other hand, just knowing the charge is not a complete description of the situation. Matter is composed of several kinds of electrically charged particles, and these particles have many properties, not just charge.

The most common charge carriers are the positively charged proton and the negatively charged electron. The movement of any of these charged particles constitutes an electric current. In many situations, it suffices to speak of the *conventional current* without regard to whether it is carried by positive charges moving in the direction of the conventional current or by negative charges moving in the opposite direction. This macroscopic viewpoint is an approximation that simplifies electromagnetic concepts and calculations.

At the opposite extreme, if one looks at the microscopic situation, one sees there are many ways of carrying an electric current, including: a flow of electrons; a flow of electron "holes" that act like positive particles; and both negative and positive particles (ions or other charged particles) flowing in opposite directions in an electrolytic solution or a plasma.

Beware that, in the common and important case of metallic wires, the direction of the conventional current is opposite to the drift velocity of the actual charge carriers; i.e., the electrons. This is a source of confusion for beginners.

ELECTRIC FIELD

An electric field is a vector field that associates to each point in space the Coulomb force that would be experienced per unit of electric charge, by an infinitesimal test charge at that point. Electric fields converge and diverge at electric charges and can be induced by time-varying magnetic fields. The electric field combines with the magnetic field to form the electromagnetic field.

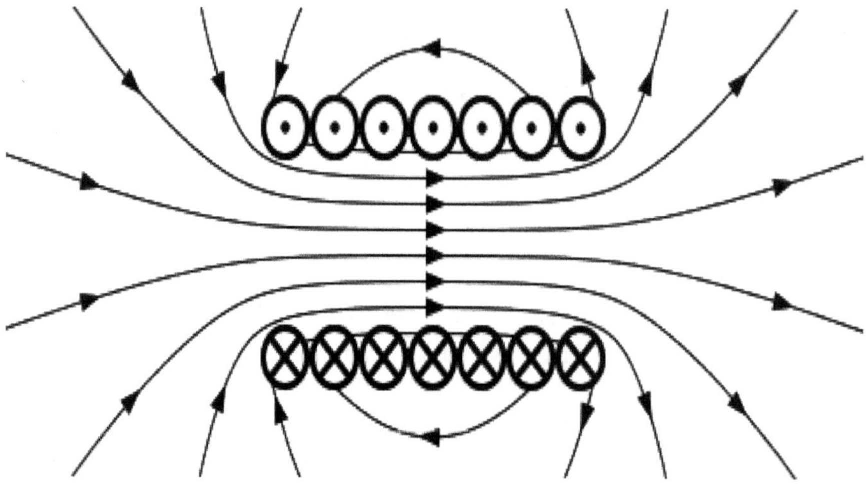

Electric field lines emanating from a point positive electric charge suspended over an infinite sheet of conducting material.

Definition

he electric field, E, at a given point is defined as the (vector) force, F, that would be exerted on a stationary test particle of unit charge by electromagnetic forces (i.e. the Lorentz force). A particle of charge q would be subject to a force F = qE.

Its SI units are newtons per coulomb ($N \cdot C^{-1}$) or, equivalently, volts per metre ($V \cdot m^{-1}$), which in terms of SI base units are $kg \cdot m \cdot s^{-3} \cdot A^{-1}$.

Sources of Electric Field

Causes and Description

Electric fields are caused by electric charges or varying magnetic fields. The former effect is described by Gauss`s law, the latter by Faraday`s law of induction, which together are enough to define the behavior of the electric field as a function of charge repartition and magnetic field. However, since the magnetic field is described as a function of electric field, the equations of both fields are coupled and together form Maxwell`s equations that describe both fields as a function of charges and currents.

In the special case of a steady state (stationary charges and currents), the Maxwell-Faraday inductive effect disappears. The resulting two equations (Gauss's law $\nabla \cdot \mathbf{E} = \frac{\rho}{\varepsilon_0}$ and Faraday's law with no induction term $\nabla \times \mathbf{E} = 0$), taken together, are equivalent to Coulomb`s law, written as $\boldsymbol{E}(\boldsymbol{r}) = \frac{1}{4\pi\varepsilon_0} \int d\boldsymbol{r}' \rho(\boldsymbol{r}') \frac{\boldsymbol{r} - \boldsymbol{r}'}{|\boldsymbol{r} - \boldsymbol{r}'|^3}$ for a charge density $\rho(\mathbf{r})$ ($\mathbf{r}$ denotes the position in space). Notice that ε_0, the permittivity of vacuum, must be substituted if charges are considered in non-empty media.

Continuous vs. Discrete Charge Repartition

The equations of electromagnetism are best described in a continuous description. However, charges are sometimes best described as discrete points; for example, some models may describe electrons as point sources where charge density is infinite on an infinitesimal section of space. A charge q located at $\mathbf{r}_0$ can be described mathematically as a charge density $\rho(\mathbf{r}) = q\delta(\mathbf{r} - \mathbf{r}_0)$, where the Dirac delta function (in three dimensions) is used. Conversely, a charge distribution can be approximated by many small point charges.

Superposition Principle

Electric fields satisfy the superposition principle, because Maxwell's equations are linear. As a result, if $\mathbf{E}_1$ and $\mathbf{E}_2$ are the electric fields resulting from distribution of charges ρ_1 and ρ_2, a distribution of charges $\rho_1 + \rho_2$ will create an electric field $\mathbf{E}_1 + \mathbf{E}_2$; for instance, Coulomb's law is linear in charge density as well.

This principle is useful to calculate the field created by multiple point charges. If charges $q_1, q_2, \ldots, q_n$ are stationary in space at $\mathbf{r}_1, \mathbf{r}_2, \ldots \mathbf{r}_n$, in the absence of currents, the superposition principle proves that the resulting field is the sum of fields generated by each particle as described by Coulomb's law:

$$\mathbf{E}(\mathbf{r}) = \sum_{i=1}^{N} \mathbf{E}_i(\mathbf{r}) = \frac{1}{4\pi\varepsilon_0} \sum_{i=1}^{N} q_i \frac{\mathbf{r} - \mathbf{r}_i}{|\mathbf{r} - \mathbf{r}_i|^3}$$

ELECTRIC POTENTIAL

An electric potential (also called the *electric field potential* or the *electrostatic potential*) is the amount of electric potential energy that a unitary point electric charge would have if located at any point in space, and is equal to the work done by an external agent in carrying a unit of positive charge from the arbitrarily chosen reference point (usually infinity) to that point without any acceleration.

According to theoretical electromagnetics, electric potential is a scalar quantity denoted by *V*, equal to the electric potential energy of any charged particle at any location (measured in joules) divided by the charge of that particle (measured in coulombs). By dividing out the charge on the particle a remainder is obtained that is a property of the electric field itself.

This value can be calculated in either a static (time-invariant) or a dynamic (varying with time) electric field at a specific time in units of joules per coulomb (J C^{-1}), or volts (V). The electric potential at infinity is assumed to be zero.

A generalized electric scalar potential is also used in electrodynamics when time-varying electromagnetic fields are present, but this can not be so simply calculated. The electric potential and the magnetic vector potential together form a four vector, so that the two kinds of potential are mixed under Lorentz transformations.

Classical mechanics explores concepts such as force, energy, potential etc. Force and potential energy are directly related. A net force acting on any object will cause it to accelerate. As an object moves in the direction in which the force accelerates it, its potential energy decreases: the gravitational potential energy of a cannonball at the top of a hill is greater than at the base of the hill. As it rolls downhill its potential energy decreases, being translated to motion, inertial (kinetic) energy.

It is possible to define the potential of certain force fields so that the potential energy of an object in that field depends only on the position of the object with respect to the field. Two such force fields are the gravitational field and an electric field (in the absence of time-varying magnetic fields). Such fields must affect objects due to the intrinsic properties of the object (e.g., mass or charge) and the position of the object.

Objects may possess a property known as electric charge and an electric field exerts a force on charged objects. If the charged object has a positive charge the force will be in the direction of the electric field vector at that point while if the charge is negative the force will be in the opposite direction. The magnitude of the force is given by the quantity of the charge multiplied by the magnitude of the electric field vector.

HISTORY

Thales, the earliest known researcher into electricity

Long before any knowledge of electricity existed, people were aware of shocks from electric fish. Ancient Egyptian texts dating from 2750 BCE referred to these fish as the "Thunderer of the Nile", and described them as the "protectors" of all other fish. Electric fish were again reported millennia later by ancient Greek, Roman and Arabic naturalists and physicians. Several ancient writers, such as Pliny the Elder and Scribonius Largus, attested to the numbing effect of electric shocks delivered by catfish and electric rays, and knew that such shocks could travel along conducting objects. Patients suffering from ailments such as gout or headache were directed to touch electric fish in the hope that the powerful jolt might cure them. Possibly the earliest and nearest approach to the discovery of the identity of lightning, and electricity from any other source, is to be attributed to the Arabs, who before the 15th century had the Arabic word for lightning (*raad*) applied to the electric ray.

Ancient cultures around the Mediterranean knew that certain objects, such as rods of amber, could be rubbed with cat`s fur to attract light objects like feathers. Thales of Miletus made a series of observations on static electricity around 600 BCE, from which he believed that friction rendered amber magnetic, in contrast to minerals such as magnetite, which needed no rubbing. Thales was incorrect in believing the attraction was due to a magnetic effect, but later science would prove a link between magnetism and electricity. According to a controversial theory, the Parthians may have had knowledge of electroplating, based on the 1936 discovery of the Baghdad Battery, which resembles a galvanic cell, though it is uncertain whether the artifact was electrical in nature.

Benjamin Franklin conducted extensive research on electricity in the 18th century, as documented by Joseph Priestley (1767) *History and Present Status of Electricity*, with whom Franklin carried on extended correspondence.

Electricity would remain little more than an intellectual curiosity for millennia until 1600, when the English scientist William Gilbert made a careful study of electricity and magnetism, distinguishing the lodestone effect from static electricity produced by rubbing amber. He coined the New Latin word *electricus* ("of amber" or "like amber", from ἤ λεκτρον, *elektron*, the Greek word for "amber") to refer to the property of attracting small objects after being rubbed. This association gave rise to the English words "electric" and "electricity", which made their first appearance in print in Thomas Browne`s *Pseudodoxia Epidemica* of 1646.

Further work was conducted by Otto von Guericke, Robert Boyle, Stephen Gray and C. F. du Fay. In the 18th century, Benjamin Franklin conducted extensive research in electricity, selling his possessions to fund his work. In June 1752 he is reputed to have attached a metal key to the bottom of a dampened kite string and flown the kite in a storm-threatened sky. A succession of sparks jumping from the key to the back of his hand showed that lightning was indeed electrical in nature. He also explained the apparently paradoxical behavior of the Leyden jar as a device for storing large amounts of electrical charge in terms of electricity consisting of both positive and negative charges.

Michael Faraday's discoveries formed the foundation of electric motor technology

In 1791, Luigi Galvani published his discovery of bioelectromagnetics, demonstrating that electricity was the medium by which neurons passed signals to the muscles. Alessandro Volta`s battery, or voltaic pile, of 1800, made from alternating layers of zinc and copper, provided scientists with a more reliable source of electrical energy than the electrostatic machines previously used. The recognition of electromagnetism, the unity of electric and magnetic phenomena, is due to Hans Christian Ørsted and André-Marie Ampère in 1819-1820; Michael Faraday invented the electric motor in 1821, and Georg Ohm mathematically analysed the electrical circuit in 1827. Electricity and magnetism (and light) were definitively linked by James Clerk Maxwell, in particular in his "On Physical Lines of Force" in 1861 and 1862.

While the early 19th century had seen rapid progress in electrical science, the late 19th century would see the greatest progress in electrical engineering. Through such people as Alexander Graham Bell, Ottó Bláthy, Thomas Edison, Galileo Ferraris, Oliver Heaviside, Ányos Jedlik, William Thomson, 1st Baron Kelvin, Charles Algernon Parsons, Werner von Siemens, Joseph Swan, Reginald Fessenden, Nikola Tesla and George Westinghouse, electricity turned from a scientific curiosity into an essential tool for modern life, becoming a driving force of the Second Industrial Revolution.

In 1887, Heinrich Hertz discovered that electrodes illuminated with ultraviolet light create electric sparks more easily. In 1905 Albert Einstein published a paper that explained experimental data from the photoelectric effect as being the result of light energy being carried in discrete quantized packets, energising electrons. This discovery led to the quantum revolution. Einstein was awarded the Nobel Prize in Physics in 1921 for "his discovery of the law of the photoelectric effect". The photoelectric effect is also employed in photocells such as can be found in solar panels and this is frequently used to make electricity commercially.

The first solid-state device was the "cat`s-whisker detector" first used in the 1900s in radio receivers. A whisker-like wire is placed lightly in contact with a solid crystal (such as a germanium crystal) in order to detect a radio signal by the contact junction effect. In a solid-state component, the current is confined to solid elements and compounds engineered specifically to switch and amplify it. Current flow can be understood in two forms: as negatively charged electrons, and as positively charged electron deficiencies called holes. These charges and holes are understood in terms of quantum physics. The building material is most often a crystalline semiconductor.

The solid-state device came into its own with thc invcntion of thc transistor in 1947. Common solid-state devices include transistors, microprocessor chips, and RAM. A specialized type of RAM called flash RAM is used in USB flash drives and more recently, solid-state drives to replace mechanically rotating magnetic disc hard disk drives. Solid state devices became prevalent in the 1950s and the 1960s, during the transition from vacuum tubes to semiconductor diodes, transistors, integrated circuit (IC) and the light-emitting diode (LED).

CONCEPTS

Electric Charge

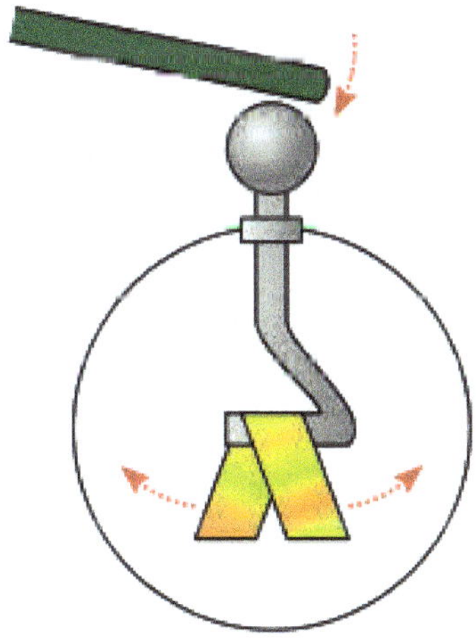

Charge on a gold-leaf electroscope causes the leaves to visibly repel each other

The presence of charge gives rise to an electrostatic force: charges exert a force on each other, an effect that was known, though not understood, in antiquity.[:457] A lightweight ball suspended from a string can be charged by touching it with a glass rod that has itself been charged by rubbing with a cloth. If a similar ball is charged by the same glass rod, it is found to repel the first: the charge acts to force the two balls apart. Two balls that are charged with a rubbed amber rod also repel each other. However, if one ball is charged by the glass rod, and the other by an amber rod, the two balls are found to attract each other. These phenomena were investigated in the late eighteenth century by Charles-Augustin de Coulomb, who deduced that charge manifests itself in two opposing forms. This discovery led to the well-known axiom: *like-charged objects repel and opposite-charged objects attract*.

The force acts on the charged particles themselves, hence charge has a tendency to spread itself as evenly as possible over a conducting surface. The magnitude of the electromagnetic force, whether attractive or repulsive, is given by Coulomb's law, which relates the force to the product of the charges and has an inverse-square relation to the distance between them.[:35] The electromagnetic force is very strong, second only in strength to the strong interaction, but unlike that force it operates over all distances. In comparison with the much weaker gravitational force, the electromagnetic force pushing two electrons apart is 10^{42} times that of the gravitational attraction pulling them together.

Study has shown that the origin of charge is from certain types of subatomic particles which have the property of electric charge. Electric charge gives rise to and interacts with the electromagnetic force, one of the four fundamental forces of nature. The most familiar carriers of electrical charge are the electron and proton. Experiment has shown charge to be a conserved quantity, that is, the net charge within an isolated system will always remain constant regardless of any changes taking place within that system. Within the system, charge may be transferred between bodies, either by direct contact, or by passing along a conducting material, such as a wire. The informal term static electricity refers to the net presence (or 'imbalance') of charge on a body, usually caused when dissimilar materials are rubbed together, transferring charge from one to the other.

The charge on electrons and protons is opposite in sign, hence an amount of charge may be expressed as being either negative or positive. By convention, the charge carried by electrons is deemed negative, and that by protons positive, a custom that originated with the work of Benjamin Franklin. The amount of charge is usually given the symbol Q and expressed in coulombs; each electron carries the same charge of approximately -1.6022×10^{-19} coulomb. The proton has a charge that is equal and opposite, and thus $+1.6022\times10^{-19}$ coulomb. Charge is possessed not just by matter, but also by antimatter, each antiparticle bearing an equal and opposite charge to its corresponding particle.

Charge can be measured by a number of means, an early instrument being the gold-leaf electroscope, which although still in use for classroom demonstrations, has been superseded by the electronic electrometer.

Electric Current

The movement of electric charge is known as an electric current, the intensity of which is usually measured in amperes. Current can consist of any moving charged particles;

most commonly these are electrons, but any charge in motion constitutes a current.

By historical convention, a positive current is defined as having the same direction of flow as any positive charge it contains, or to flow from the most positive part of a circuit to the most negative part. Current defined in this manner is called conventional current. The motion of negatively charged electrons around an electric circuit, one of the most familiar forms of current, is thus deemed positive in the *opposite* direction to that of the electrons. However, depending on the conditions, an electric current can consist of a flow of charged particles in either direction, or even in both directions at once. The positive-to-negative convention is widely used to simplify this situation.

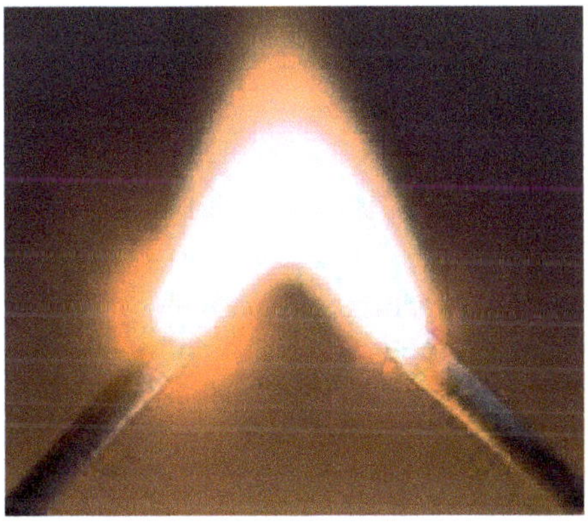

An electric arc provides an energetic demonstration of electric current

The process by which electric current passes through a material is termed electrical conduction, and its nature varies with that of the charged particles and the material through which they are travelling. Examples of electric currents include metallic conduction, where electrons flow through a conductor such as metal, and electrolysis, where ions (charged atoms) flow through liquids, or through plasmas such as electrical sparks. While the particles themselves can move quite slowly, sometimes with an average drift velocity only fractions of a millimetre per second,[17] the electric field that drives them itself propagates at close to the speed of light, enabling electrical signals to pass rapidly along wires.

Current causes several observable effects, which historically were the means of recognising its presence. That water could be decomposed by the current from a voltaic pile was discovered by Nicholson and Carlisle in 1800, a process now known as electrolysis. Their work was greatly expanded upon by Michael Faraday in 1833. Current through a resistance causes localised heating, an effect James Prescott Joule studied mathematically in 1840. One of the most important discoveries relating to current was made accidentally by Hans Christian Ørsted in 1820, when, while preparing a lecture, he witnessed the current in a wire disturbing the needle of a magnetic compass. He had discovered electromagnetism, a fundamental interaction between electricity and magnetics. The level of electromagnetic emissions generated by electric arcing is high enough to produce electromagnetic interference, which can be detrimental to the workings of adjacent equipment.

In engineering or household applications, current is often described as being either direct current (DC) or alternating current (AC). These terms refer to how the current varies in time. Direct current, as produced by example from a battery and required by most electronic devices, is a unidirectional flow from the positive part of a circuit to the negative.[11] If, as is most common, this flow is carried by electrons, they will be travelling in the opposite direction. Alternating current is any current that reverses direction repeatedly; almost always this takes the form of a sine wave. Alternating current thus pulses back and forth within a conductor without the charge moving any net distance over time. The time-averaged value of an alternating current is zero, but it delivers energy in first one direction, and then the reverse. Alternating current is affected by electrical properties that are not observed under steady state direct current, such as inductance and capacitance. These properties however can become important when circuitry is subjected to transients, such as when first energised.

Electric Field

The concept of the electric field was introduced by Michael Faraday. An electric field is created by a charged body in the space that surrounds it, and results in a force exerted on any other charges placed within the field. The electric field acts between two charges in a similar manner to the way that the gravitational field acts between two masses, and like it, extends towards infinity and shows an inverse square relationship with distance. However, there is an important difference. Gravity always acts in attraction, drawing two masses together, while the electric field can result in either attraction or repulsion. Since large bodies such as planets generally carry no net charge, the electric field at a distance is usually zero. Thus gravity is the dominant force at distance in the universe, despite being much weaker.

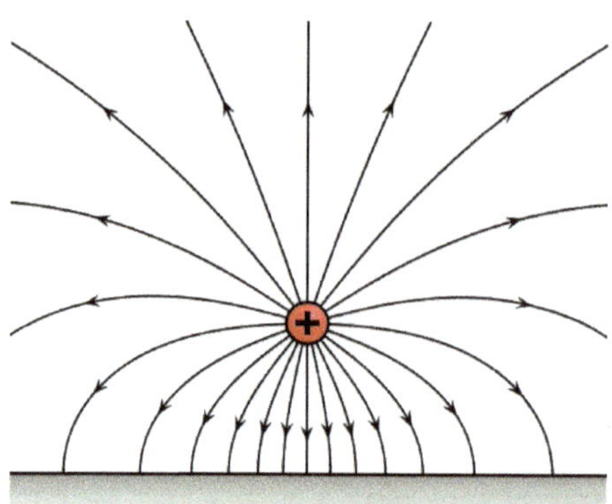

Field lines emanating from a positive charge above a plane conductor

An electric field generally varies in space, and its strength at any one point is defined as the force (per unit charge) that would be felt by a stationary, negligible charge if placed at that point. The conceptual charge, termed a 'test charge', must be vanishingly small to prevent its own electric field disturbing the main field and must also be stationary to prevent the effect of magnetic fields. As the electric field is defined in terms of force, and force is a vector, so it follows that an electric field is also a vector, having both magnitude and direction. Specifically, it is a vector field.

The study of electric fields created by stationary charges is called electrostatics. The field may be visualised by a set of imaginary lines whose direction at any point is the same as that of the field. This concept was introduced by Faraday, whose term 'lines of force' still sometimes sees use. The field lines are the paths that a point positive charge would seek to make as it was forced to move within the field; they are however an imaginary concept with no physical existence, and the field permeates all the intervening space between the lines. Field lines emanating from stationary charges have several key properties: first, that they originate at positive charges and terminate at negative charges; second, that they must enter any good conductor at right angles, and third, that they may never cross nor close in on themselves.

A hollow conducting body carries all its charge on its outer surface. The field is therefore zero at all places inside the body.[88] This is the operating principal of the Faraday cage, a conducting metal shell which isolates its interior from outside electrical effects.

The principles of electrostatics are important when designing items of high-voltage equipment. There is a finite limit to the electric field strength that may be withstood by any medium. Beyond this point, electrical breakdown occurs and an electric arc causes flashover between the charged parts. Air, for example, tends to arc across small gaps at electric field strengths which exceed 30 kV per centimetre. Over larger gaps, its breakdown strength is weaker, perhaps 1 kV per centimetre. The most visible natural occurrence of this is lightning, caused when charge becomes separated in the clouds by rising columns of air, and raises the electric field in the air to greater than it can withstand. The voltage of a large lightning cloud may be as high as 100 MV and have discharge energies as great as 250 kWh.

The field strength is greatly affected by nearby conducting objects, and it is particularly intense when it is forced to curve around sharply pointed objects. This principle is exploited in the lightning conductor, the sharp spike of which acts to encourage the lightning stroke to develop there, rather than to the building it serves to protect

Electric Potential

A pair of AA cells. The + sign indicates the polarity of the potential difference between the battery terminals.

The concept of electric potential is closely linked to that of the electric field. A small charge placed within an electric field experiences a force, and to have brought that charge to that point against the force requires work. The electric potential at any point is defined as the energy required to bring a unit test charge from an infinite distance slowly to that point. It is usually measured in volts, and one volt is the potential for which one joule of work must be expended to bring a charge of one coulomb from infinity. This definition of potential, while formal, has little practical application, and a more useful concept is that of electric potential difference, and is the energy required to move a unit charge between two specified points. An electric field has the special property that it is *conservative*, which means that the path taken by the test charge is irrelevant: all paths between two specified points expend the same energy, and thus a unique value for potential difference may be stated. The volt is so strongly identified as the unit of choice for measurement and description of electric potential difference that the term voltage sees greater everyday usage.

For practical purposes, it is useful to define a common reference point to which potentials may be expressed and compared. While this could be at infinity, a much more useful reference is the Earth itself, which is assumed to be at the same potential everywhere. This reference point naturally takes the name earth or ground. Earth is assumed to be an infinite source of equal amounts of positive and negative charge, and is therefore electrically uncharged—and unchargeable.

Electric potential is a scalar quantity, that is, it has only magnitude and not direction. It may be viewed as analogous to height: just as a released object will fall through a difference in heights caused by a gravitational field, so a charge will ‹fall` across the voltage caused by an electric field.As relief maps show contour lines marking points of equal height, a set of lines marking points of equal potential (known as equipotentials) may be drawn around an electrostatically charged object. The equipotentials cross all lines of force at right angles. They must also lie parallel to a conductor`s surface, otherwise this would produce a force that will move the charge carriers to even the potential of the surface.

The electric field was formally defined as the force exerted per unit charge, but the concept of potential allows for a more useful and equivalent definition: the electric field is the local gradient of the electric potential. Usually expressed in volts per metre, the vector direction of the field is the line of greatest slope of potential, and where the equipotentials lie closest together.

Electromagnets

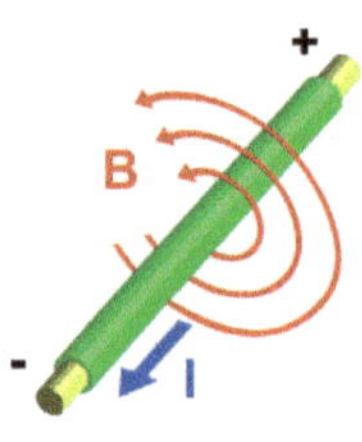

Magnetic field circles around a current

Ørsted's discovery in 1821 that a magnetic field existed around all sides of a wire carrying an electric current indicated that there was a direct relationship between electricity and magnetism. Moreover, the interaction seemed different from gravitational and electrostatic forces, the two forces of nature then known. The force on the compass needle did not direct it to or away from the current-carrying wire, but acted at right angles to it. Ørsted's slightly obscure words were that "the electric conflict acts in a revolving manner." The force also depended on the direction of the current, for if the flow was reversed, then the force did too.

Ørsted did not fully understand his discovery, but he observed the effect was reciprocal: a current exerts a force on a magnet, and a magnetic field exerts a force on a current. The phenomenon was further investigated by Ampère, who discovered that two parallel current-carrying wires exerted a force upon each other: two wires conducting currents in the same direction are attracted to each other, while wires containing currents in opposite directions are forced apart. The interaction is mediated by the magnetic field each current produces and forms the basis for the international definition of the ampere.

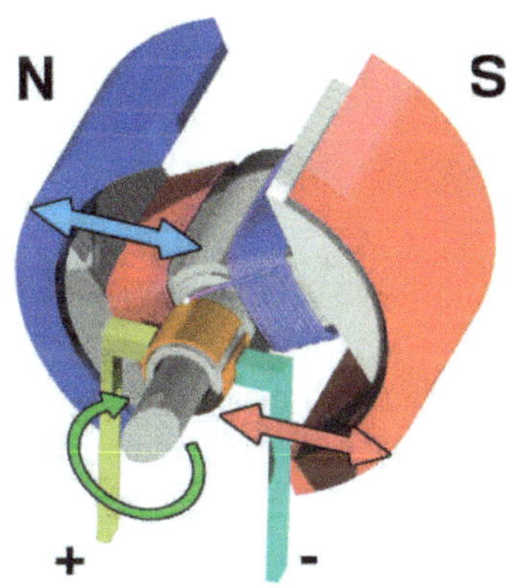

The electric motor exploits an important effect of electromagnetism: a current through a magnetic field experiences a force at right angles to both the field and current

This relationship between magnetic fields and currents is extremely important, for it led to Michael Faraday's invention of the electric motor in 1821. Faraday's homopolar motor consisted of a permanent magnet sitting in a pool of mercury. A current was allowed through a wire suspended from a pivot above the magnet and dipped into the mercury. The magnet exerted a tangential force on the wire, making it circle around the magnet for as long as the current was maintained.

Experimentation by Faraday in 1831 revealed that a wire moving perpendicular to a magnetic field developed a potential difference between its ends. Further analysis of this process, known as electromagnetic induction, enabled him to state the principle, now known as Faraday`s law of induction, that the potential difference induced in a closed circuit is proportional to the rate of change of magnetic flux through the loop. Exploitation of this discovery enabled him to invent the first electrical generator in 1831, in which he converted the mechanical energy of a rotating copper disc to electrical energy. Faraday`s disc was inefficient and of no use as a practical generator, but it showed the possibility of generating electric power using magnetism, a possibility that would be taken up by those that followed on from his work.

Electrochemistry

Italian physicist Alessandro Volta showing his *«battery"* to Frenchemperor Napoleon Bonaparte in the early 19th century.

The ability of chemical reactions to produce electricity, and conversely the ability of electricity to drive chemical reactions has a wide array of uses.

Electrochemistry has always been an important part of electricity. From the initial invention of the Voltaic pile, electrochemical cells have evolved into the many different types of batteries, electroplating and electrolysis cells. Aluminium is produced in vast quantities this way, and many portable devices are electrically powered using rechargeable cells.

Electric Circuits

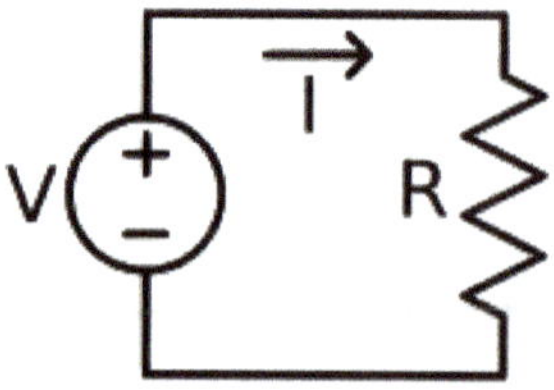

A basic electric circuit. The voltage source *V* on the left drives a current *I* around the circuit, delivering electrical energy into the resistor *R*. From the resistor, the current returns to the source, completing the circuit.

An electric circuit is an interconnection of electric components such that electric charge is made to flow along a closed path (a circuit), usually to perform some useful task.

The components in an electric circuit can take many forms, which can include elements such as resistors, capacitors, switches, transformers and electronics. Electronic circuits contain active components, usually semiconductors, and typically exhibit non-linear behaviour, requiring complex analysis. The simplest electric components are those that are termed passive and linear: while they may temporarily store energy, they contain no sources of it, and exhibit linear responses to stimuli.

The resistor is perhaps the simplest of passive circuit elements: as its name suggests, it resists the current through it, dissipating its energy as heat. The resistance is a consequence of the motion of charge through a conductor: in metals, for example, resistance is primarily due to collisions between electrons and ions. Ohm`s law is a basic law of circuit theory, stating that the current passing through a resistance is directly proportional to the potential difference across it. The resistance of most materials is relatively constant over a range of temperatures and currents; materials under these conditions are known as ‹ohmic`. The ohm, the unit of resistance, was named in honour of Georg Ohm, and is symbolised by the Greek letter Ω. 1 Ω is the resistance that will produce a potential difference of one volt in response to a current of one amp.

The capacitor is a development of the Leyden jar and is a device that can store charge, and thereby storing electrical energy in the resulting field. It consists of two conducting plates separated by a thin insulating dielectric layer; in practice, thin metal foils are coiled together, increasing the surface area per unit volume and therefore the capacitance. The unit of capacitance is the farad, named after Michael Faraday, and given the symbol *F*: one farad is the capacitance that develops a potential difference of one volt when it stores a charge of one coulomb. A capacitor connected to a voltage supply initially causes a current as it accumulates charge; this current will however decay in time as the capacitor fills, eventually falling to zero. A capacitor will therefore not permit a steady state current, but instead blocks it.

The inductor is a conductor, usually a coil of wire, that stores energy in a magnetic field in response to the current through it. When the current changes, the magnetic field does too, inducing a voltage between the ends of the conductor. The induced voltage is proportional to the time rate of change of the current. The constant of proportionality is termed the inductance. The unit of inductance is the henry, named after Joseph Henry, a contemporary of Faraday. One henry is the inductance that will induce a potential difference of one volt if the current through it changes at a rate of one ampere per second. The inductor`s behaviour is in some regards converse to that of the capacitor: it will freely allow an unchanging current, but opposes a rapidly changing one.

Electric Power

Electric power is the rate at which electric energy is transferred by an electric circuit. The SI unit of power is the watt, one joule per second.

Electric power, like mechanical power, is the rate of doing work, measured in watts, and represented by the letter *P*. The term *wattage* is used colloquially to mean "electric power in watts." The electric power in watts produced by an electric current I consisting of a charge of Q coulombs every t seconds passing through an electric potential (voltage) difference of V is

$$P = \text{work done per unit time} = \frac{QV}{t} = IV$$

where

Q is electric charge in coulombs

t is time in seconds

I is electric current in amperes

V is electric potential or voltage in volts

Electricity generation is often done with electric generators, but can also be supplied by chemical sources such as electric batteries or by other means from a wide variety of sources of energy. Electric power is generally supplied to businesses and homes by the electric power industry. Electricity is usually sold by the kilowatt hour (3.6 MJ) which is the product of power in kilowatts multiplied by running time in hours. Electric utilities measure power using electricity meters, which keep a running total of the electric energy delivered to a customer. Unlike fossil fuels, electricity is a low entropy form of energy and can be converted into motion or many other forms of energy with high efficiency.

Electronics

Surface mount electronic components

Electronics deals with electrical circuits that involve active electrical components such as vacuum tubes, transistors, diodes and integrated circuits, and associated passive interconnection technologies. The nonlinear behaviour of active components and their ability to control electron flows makes amplification of weak signals possible and electronics is widely used in information processing, telecommunications, and signal processing. The ability of electronic devices to act as switches makes digital information processing possible. Interconnection technologies such as circuit boards, electronics packaging technology, and other varied forms of communication infrastructure complete circuit functionality and transform the mixed components into a regular working system.

Today, most electronic devices use semiconductor components to perform electron control. The study of semiconductor devices and related technology is considered a branch of solid state physics, whereas the design and construction of electronic circuits to solve practical problems come under electronics engineering.

Electromagnetic Wave

Faraday's and Ampère's work showed that a time-varying magnetic field acted as a source of an electric field, and a time-varying electric field was a source of a magnetic field. Thus, when either field is changing in time, then a field of the other is necessarily induced. Such a phenomenon has the properties of a wave, and is naturally referred to as an electromagnetic wave. Electromagnetic waves were analysed theoretically by James Clerk Maxwell in 1864. Maxwell developed a set of equations that could unambiguously describe the interrelationship between electric field, magnetic field, electric charge, and electric current. He could moreover prove that such a wave would necessarily travel at the speed of light, and thus light itself was a form of electromagnetic radiation. Maxwell`s Laws, which unify light, fields, and charge are one of the great milestones of theoretical physics.

Thus, the work of many researchers enabled the use of electronics to convert signals into high frequency oscillating currents, and via suitably shaped conductors, electricity permits the transmission and reception of these signals via radio waves over very long distances.

Production and Uses

Generation and Transmission

Early 20th-century alternator made in Budapest, Hungary, in the power generating hall of a hydroelectric station (photograph by Prokudin-Gorsky, 1905–1915).

In the 6th century BC, the Greek philosopher Thales of Miletus experimented with amber rods and these experiments were the first studies into the production of electrical energy. While this method, now known as the triboelectric effect, can lift light objects and generate sparks, it is extremely inefficient. It was not until the invention of the voltaic pile in the eighteenth century that a viable source of electricity became available. The voltaic pile, and its modern descendant, the electrical battery, store energy chemically and make it available on demand in the form of electrical energy. The battery is a versatile and very common power source which is ideally suited to many applications, but its energy storage is finite, and once discharged it must be disposed of or recharged. For large electrical demands electrical energy must be generated and transmitted continuously over conductive transmission lines.

Electrical power is usually generated by electro-mechanical generators driven by steam produced from fossil fuel combustion, or the heat released from nuclear reactions; or from other sources such as kinetic energy extracted from wind or flowing water. The modern steam turbine invented by Sir Charles Parsons in 1884 today generates about 80 percent of the electric power in the world using a variety of heat sources. Such generators bear no resemblance to Faraday's homopolar disc generator of 1831, but they still rely on his electromagnetic principle that a conductor linking a changing magnetic field induces a potential difference across its ends. The invention in the late nineteenth century of the transformer meant that electrical power could be transmitted more efficiently at a higher voltage but lower current. Efficient electrical transmission meant in turn that electricity could be generated at centralised power stations, where it benefited from economies of scale, and then be despatched relatively long distances to where it was needed.

Wind power is of increasing importance in many countries

Since electrical energy cannot easily be stored in quantities large enough to meet demands on a national scale, at all times exactly as much must be produced as is required. This requires electricity utilities to make careful predictions of their electrical loads, and maintain constant coordination with their power stations. A certain amount of generation must always be held in reserve to cushion an electrical grid against inevitable disturbances and losses.

Demand for electricity grows with great rapidity as a nation modernizes and its economy develops. The United States showed a 12% increase in demand during each year of the first three decades of the twentieth century, a rate of growth that is now being experienced by emerging economies such as those of India or China. Historically, the growth rate for electricity demand has outstripped that for other forms of energy.

Environmental concerns with electricity generation have led to an increased focus on generation from renewable sources, in particular from wind and hydropower. While debate can be expected to continue over the environmental impact of different means of electricity production, its final form is relatively clean

Applications

The light bulb, an early application of electricity, operates by Joule heating: the passage of current through resistance generating heat

Electricity is a very convenient way to transfer energy, and it has been adapted to a huge, and growing, number of uses. The invention of a practical incandescent light bulb in the 1870s led to lighting becoming one of the first publicly available applications of electrical power. Although electrification brought with it its own dangers, replacing the naked flames of gas lighting greatly reduced fire hazards within homes and factories. Public utilities were set up in many cities targeting the burgeoning market for electrical lighting.

The resistive Joule heating effect employed in filament light bulbs also sees more direct use in electric heating. While this is versatile and controllable, it can be seen as wasteful, since most electrical generation has already required the production of heat at a power station. A number of countries, such as Denmark, have issued legislation restricting or banning the use of resistive electric heating in new buildings. Electricity is however still a highly practical energy source for heating and refrigeration, with air conditioning/ heat pumps representing a growing sector for electricity demand for heating and cooling, the effects of which electricity utilities are increasingly obliged to accommodate.

Electricity is used within telecommunications, and indeed the electrical telegraph, demonstrated commercially in 1837 by Cooke and Wheatstone, was one of its earliest applications. With the construction of first intercontinental, and then transatlantic, telegraph systems in the 1860s, electricity had enabled communications in minutes across the globe. Optical fibre and satellite communication have taken a share of the market for communications systems, but electricity can be expected to remain an essential part of the process.

The effects of electromagnetism are most visibly employed in the electric motor, which provides a clean and efficient means of motive power. A stationary motor such as a winch is easily provided with a supply of power, but a motor that moves with its application, such as an electric vehicle, is obliged to either carry along a power source such as a battery, or to collect current from a sliding contact such as a pantograph.

Electronic devices make use of the transistor, perhaps one of the most important inventions of the twentieth century, and a fundamental building block of all modern circuitry. A modern integrated circuit may contain several billion miniaturised transistors in a region only a few centimetres square.

Electricity is also used to fuel public transportation, including electric buses and trains.

ELECTRICITY AND THE NATURAL WORLD

Physiological Effects

A voltage applied to a human body causes an electric current through the tissues, and although the relationship is non-linear, the greater the voltage, the greater the current. The threshold for perception varies with the supply frequency and with the path of the current, but is about 0.1 mA to 1 mA for mains-frequency electricity, though a current as low as a microamp can be detected as an electrovibration effect under certain conditions. If the current is sufficiently high, it will cause muscle contraction, fibrillation of the heart, and tissue burns. The lack of any visible sign that a conductor is electrified makes electricity a particular hazard. The pain caused by an electric shock can be intense, leading electricity at times to be employed as a method of torture. Death caused by an electric shock is referred to as electrocution. Electrocution is still the means of judicial execution in some jurisdictions, though its use has become rarer in recent times.

Electrical Phenomena in Nature

The electric eel, *Electrophorus electricus*

Electricity is not a human invention, and may be observed in several forms in nature, a prominent manifestation of which is lightning. Many interactions familiar at the macroscopic level, such as touch, friction or chemical bonding, are due to interactions between electric fields on the atomic scale. The Earth`s magnetic field is thought to arise from a natural dynamo of circulating currents in the planet's core.Certain crystals, such as quartz, or even sugar, generate a potential difference across their faces when subjected to external pressure. This phenomenon is known as piezoelectricity, from the Greek *piezein* (πιέζειν), meaning to press, and was discovered in 1880 by Pierre and Jacques Curie. The effect is reciprocal, and when a piezoelectric material is subjected to an electric field, a small change in physical dimensions takes place.

Some organisms, such as sharks, are able to detect and respond to changes in electric fields, an ability known as electroreception, while others, termed electrogenic, are able to generate voltages themselves to serve as a predatory or defensive weapon. The order Gymnotiformes, of which the best known example is the electric eel, detect or stun their prey via high voltages generated from modified muscle cells called electrocytes. All animals transmit information along their cell membranes with voltage pulses called action potentials, whose functions include communication by the nervous system between neurons and muscles. An electric shock stimulates this system, and causes muscles to contract. Action potentials are also responsible for coordinating activities in certain plants.

Cultural Perception

In 1850, William Gladstone asked the scientist Michael Faraday why electricity was valuable. Faraday answered, "One day sir, you may tax it."

In the 19th and early 20th century, electricity was not part of the everyday life of many people, even in the industrialised Western world. The popular culture of the time accordingly often depicts it as a mysterious, quasi-magical force that can slay the living, revive the dead or otherwise bend the laws of nature. This attitude began with the 1771 experiments of Luigi Galvani in which the legs of dead frogs were shown to twitch on application of animal electricity. "Revitalization" or resuscitation of apparently dead or drowned persons was reported in the medical literature shortly after Galvani`s work. These results were known to Mary Shelley when she authored *Frankenstein* (1819), although she does not name the method of revitalization of the monster. The revitalization of monsters with electricity later became a stock theme in horror films.

As the public familiarity with electricity as the lifeblood of the Second Industrial Revolution grew, its wielders were more often cast in a positive light, such as the workers who "finger death at their gloves' end as they piece and repiece the living wires" in Rudyard Kipling`s 1907 poem *Sons of Martha*. Electrically powered vehicles of every sort featured large in adventure stories such as those of Jules Verne and the *Tom Swift* books. The masters of electricity, whether fictional or real—including scientists such as Thomas Edison, Charles Steinmetz or Nikola Tesla—were popularly conceived of as having wizard-like powers.

With electricity ceasing to be a novelty and becoming a necessity of everyday life in the later half of the 20th century, it required particular attention by popular culture only when it *stops* flowing, an event that usually signals disaster. The people who *keep* it flowing, such as the nameless hero of Jimmy Webb's song "Wichita Lineman" (1968), are still often cast as heroic, wizard-like figures.

REFERENCES

1. Jones, D.A. (1991), "Electrical engineering: the backbone of society", Proceedings of the IEE: Science, Measurement and Technology, 138 (1): 1–10, doi:10.1049/ip-a-3.1991.0001
2. Moller, Peter; Kramer, Bernd (December 1991), "Review: Electric Fish", BioScience, American Institute of Biological Sciences, 41 (11): 794–6 , doi:10.2307/1311732, JSTOR 1311732
3. Bullock, Theodore H. (2005), Electroreception, Springer, pp. 5–7, ISBN 0-387-23192-7
4. Morris, Simon C. (2003), Life's Solution: Inevitable Humans in a Lonely Universe, Cambridge University Press, pp. 182–185, ISBN 0-521-82704-3
5. The Encyclopedia Americana; a library of universal knowledge (1918), New York: Encyclopedia Americana Corp
6. Stewart, Joseph (2001), Intermediate Electromagnetic Theory, World Scientific, p. 50, ISBN 981-02-4471-1
7. Simpson, Brian (2003), Electrical Stimulation and the Relief of Pain, Elsevier Health Sciences, pp. 6–7, ISBN 0-444-51258-6
8. Frood, Arran (27 February 2003), Riddle of 'Baghdad's batteries', BBC, retrieved 2008-02-16
9. Baigrie, Brian (2006), Electricity and Magnetism: A Historical Perspective, Greenwood Press, pp. 7–8, ISBN 0-313-33358-0
10. Chalmers, Gordon (1937), "The Lodestone and the Understanding of Matter

in Seventeenth Century England", Philosophy of Science, 4 (1): 75–95, doi:10.1086/286445

11. Srodes, James (2002), Franklin: The Essential Founding Father, Regnery Publishing, pp. 92–94, ISBN 0-89526-163-4 It is uncertain if Franklin personally carried out this experiment, but it is popularly attributed to him.
12. Uman, Martin (1987), All About Lightning (PDF), Dover Publications, ISBN 0-486-25237-X
13. Riskin, Jessica (1998), Poor Richard's Leyden Jar: Electricity and economy in Franklinist France (PDF), p. 327
14. Kirby, Richard S. (1990), Engineering in History, Courier Dover Publications, pp. 331–333, ISBN 0-486-26412-2
15. Berkson, William (1974) Fields of force: the development of a world view from Faraday to Einstein p.148. Routledge, 1974
16. Marković, Dragana, The Second Industrial Revolution, archived from the original on 2007-11-19, retrieved 2007-12-09
17. Sears, Francis; et al. (1982), University Physics, Sixth Edition, Addison Wesley, ISBN 0-201-07199-1
18. Hertz, Heinrich (1887). "Ueber den Einfluss des ultravioletten Lichtes auf die electrische Entladung". Annalen der Physik. 267 (8): S. 983–1000. Bibcode:1887AnP...267..983H. doi:10.1002/andp.18872670827.
19. "The Nobel Prize in Physics 1921". Nobel Foundation. Retrieved 2013-03-16.
20. "Solid state", The Free Dictionary
21. John Sydney Blakemore, Solid state physics, pp.1-3, Cambridge University Press, 1985 ISBN 0-521-31391-0.
22. Richard C. Jaeger, Travis N. Blalock, Microelectronic circuit design, pp.46-47, McGraw-Hill Professional, 2003 ISBN 0-07-250503-6.
23. "The repulsive force between two small spheres charged with the same type of electricity is inversely proportional to the square of the distance between the centres of the two spheres." Charles-Augustin de Coulomb, Histoire de l'Academie Royal des Sciences, Paris 1785.
24. Duffin, W.J. (1980), Electricity and Magnetism, 3rd edition, McGraw-Hill, ISBN 0-07-084111-X
25. National Research Council (1998), Physics Through the 1990s, National Academies Press, pp. 215–216, ISBN 0-309-03576-7
26. Umashankar, Korada (1989), Introduction to Engineering Electromagnetic Fields, World Scientific, pp. 77–79, ISBN 9971-5-0921-0
27. Hawking, Stephen (1988), A Brief History of Time, Bantam Press, p. 77, ISBN 0-553-17521-1
28. Trefil, James (2003), The Nature of Science: An A–Z Guide to the Laws and Principles Governing Our Universe, Houghton Mifflin Books, p. 74, ISBN 0-618-31938-7

29. Shectman, Jonathan (2003), Groundbreaking Scientific Experiments, Inventions, and Discoveries of the 18th Century, Greenwood Press, pp. 87–91, ISBN 0-313-32015-2

30. Sewell, Tyson (1902), The Elements of Electrical Engineering, Lockwood, p. 18. The Q originally stood for 'quantity of electricity', the term 'electricity' now more commonly expressed as 'charge'.

31. Close, Frank (2007), The New Cosmic Onion: Quarks and the Nature of the Universe, CRC Press, p. 51, ISBN 1-58488-798-2

32. Ward, Robert (1960), Introduction to Electrical Engineering, Prentice-Hall, p. 18

33. Solymar, L. (1984), Lectures on electromagnetic theory, Oxford University Press, p. 140, ISBN 0-19-856169-5

34. Berkson, William (1974), Fields of Force: The Development of a World View from Faraday to Einstein, Routledge, p. 370, ISBN 0-7100-7626-6 Accounts differ as to whether this was before, during, or after a lecture.

35. "Lab Note #105 EMI Reduction - Unsuppressed vs. Suppressed". Arc Suppression Technologies. April 2011. Retrieved March 7, 2012.

36. Bird, John (2007), Electrical and Electronic Principles and Technology, 3rd edition, Newnes, ISBN 9781417505432

37. Almost all electric fields vary in space. An exception is the electric field surrounding a planar conductor of infinite extent, the field of which is uniform.

38. Morely & Hughes, Principles of Electricity, Fifth edition, p. 73, ISBN 0-582-42629-4

39. Naidu, M.S.; Kamataru, V. (1982), High Voltage Engineering, Tata McGraw-Hill, p. 2, ISBN 0-07-451786-4

40. Naidu, M.S.; Kamataru, V. (1982), High Voltage Engineering, Tata McGraw-Hill, pp. 201–202, ISBN 0-07-451786-4

41. Paul J. Nahin (9 October 2002). Oliver Heaviside: The Life, Work, and Times of an Electrical Genius of the Victorian Age. JHU Press. ISBN 978-0-8018-6909-9.

42. Serway, Raymond A. (2006), Serway's College Physics, Thomson Brooks, p. 500, ISBN 0-534-99724-4

43. Saeli, Sue; MacIsaac, Dan (2007), "Using Gravitational Analogies To Introduce Elementary Electrical Field Theory Concepts", The Physics Teacher, 45 (2): 104, Bibcode:2007PhTea..45..104S, doi:10.1119/1.2432088, retrieved 2007-12-09

44. Thompson, Silvanus P. (2004), Michael Faraday: His Life and Work, Elibron Classics, p. 79, ISBN 1-4212-7387-X

45. Morely & Hughes, Principles of Electricity, Fifth edition, pp. 92–93

46. Institution of Engineering and Technology, Michael Faraday: Biography, archived from the original on 2007-07-03, retrieved 2007-12-09

47. Alexander, Charles; Sadiku, Matthew (2006), Fundamentals of Electric Circuits (3, revised ed.), McGraw-Hill, ISBN 9780073301150

48. Environmental Physics By Clare Smith 2001

49. Dell, Ronald; Rand, David (2001), "Understanding Batteries", Unknown, Royal Society of Chemistry, 86: 2–4, Bibcode:1985STIN...8619754M, ISBN 0-85404-605-4
50. McLaren, Peter G. (1984), Elementary Electric Power and Machines, Ellis Horwood, pp. 182–183, ISBN 0-85312-269-5
51. Patterson, Walter C. (1999), Transforming Electricity: The Coming Generation of Change, Earthscan, pp. 44–48, ISBN 1-85383-341-X
52. Edison Electric Institute, History of the Electric Power Industry, archived from the original on November 13, 2007, retrieved 2007-12-08
53. Edison Electric Institute, History of the U.S. Electric Power Industry, 1882-1991, retrieved 2007-12-08
54. Carbon Sequestration Leadership Forum, An Energy Summary of India, archived from the original on 2007-12-05, retrieved 2007-12-08
55. IndexMundi, China Electricity - consumption, retrieved 2007-12-08
56. National Research Council (1986), Electricity in Economic Growth, National Academies Press, ISBN 0-309-03677-1
57. Wald, Matthew (21 March 1990), "Growing Use of Electricity Raises Questions on Supply", New York Times, retrieved 2007-12-09
58. d'Alroy Jones, Peter, The Consumer Society: A History of American Capitalism, Penguin Books, p. 211
59. ReVelle, Charles and Penelope (1992), The Global Environment: Securing a Sustainable Future, Jones & Bartlett, p. 298, ISBN 0-86720-321-8
60. Danish Ministry of Environment and Energy, "F.2 The Heat Supply Act", Denmark's Second National Communication on Climate Change, archived from the original on January 8, 2008, retrieved 2007-12-09
61. Brown, Charles E. (2002), Power resources, Springer, ISBN 3-540-42634-5
62. Hojjati, B.; Battles, S., The Growth in Electricity Demand in U.S. Households, 1981-2001: Implications for Carbon Emissions (PDF), retrieved 2007-12-09
63. Herrick, Dennis F. (2003), Media Management in the Age of Giants: Business Dynamics of Journalism, Blackwell Publishing, ISBN 0-8138-1699-8
64. Das, Saswato R. (2007-12-15), "The tiny, mighty transistor", Los Angeles Times
65. "Public Transportation", Alternative Energy News, 2010-03-10
66. Tleis, Nasser (2008), Power System Modelling and Fault Analysis, Elsevier, pp. 552–554, ISBN 978-0-7506-8074-5
67. Grimnes, Sverre (2000), Bioimpedance and Bioelectricity Basic, Academic Press, pp. 301–309, ISBN 0-12-303260-1
68. Lipschultz, J.H.; Hilt, M.L.J.H. (2002), Crime and Local Television News, Lawrence Erlbaum Associates, p. 95, ISBN 0-8058-3620-9
69. Encrenaz, Thérèse (2004), The Solar System, Springer, p. 217, ISBN 3-540-00241-3

70. Lima-de-Faria, José; Buerger, Martin J. (1990), Historical Atlas of Crystallography, Springer, p. 67, ISBN 0-7923-0649-X
71. Ivancevic, Vladimir & Tijana (2005), Natural Biodynamics, World Scientific, p. 602, ISBN 981-256-534-5
72. Kandel, E.; Schwartz, J.; Jessell, T. (2000), Principles of Neural Science, McGraw-Hill Professional, pp. 27–28, ISBN 0-8385-7701-6
73. Davidovits, Paul (2007), Physics in Biology and Medicine, Academic Press, pp. 204–205, ISBN 978-0-12-369411-9
74. Jackson, Mark (4 November 2013), Theoretical physics – like sex, but with no need to experiment, The Conversation
75. Van Riper, A. Bowdoin (2002), Science in popular culture: a reference guide, Westport: Greenwood Press, p. 69, ISBN 0-313-31822-0
76. Van Riper, op.cit., p. 71.

CHAPTER 2

Developments in Electrical Engineering

Chapter Contents

Electrical engineering (sometimes referred to as electrical and electronic engineering) is a professional engineering discipline that deals with the study and application of electricity, electronics and electromagnetism.

INTRODUCTION

Electrical engineering is a field of engineering that generally deals with the study and application of electricity, electronics, and electromagnetism. This field first became an identifiable occupation in the later half of the 19th century after commercialization of the electric telegraph, the telephone, and electric power distribution and use. Subsequently, broadcasting and recording media made electronics part of daily life. The invention of the transistor, and later the integrated circuit, brought down the cost of electronics to the point they can be used in almost any household object.

Electrical engineering has now subdivided into a wide range of subfields including electronics, digital computers, power engineering, telecommunications, control systems, radio-frequency engineering, signal processing, instrumentation, and microelectronics. Many of these subdisciplines overlap and also overlap with other engineering branches, spanning a huge number of specializations such as hardware engineering, power electronics, electromagnetics & waves, microwave engineering, nanotechnology, electrochemistry, renewable energies, mechatronics, electrical materials science, and many more.

Electrical engineers typically hold a degree in electrical engineering or electronic engineering. Practicing engineers may have professional certification and be members of a professional body. Such bodies include the Institute of Electrical and Electronics Engineers (IEEE) and the Institution of Engineering and Technology (IET).

Electrical engineers work in a very wide range of industries and the skills required are likewise variable. These range from basic circuit theory to the management skills required of a project manager. The tools and equipment that an individual engineer may need are similarly variable, ranging from a simple voltmeter to a top end analyzer to sophisticated design and manufacturing software.

ENIAC in Philadelphia as Glen Beck (background) and Betty Snyder (foreground) program it in BRL building 328. Photo circa. 1947 to 1955

Electrical engineers design complex power systems on a macroscopic level as well as microscopic electronic devices

Smaller length scale control centers

and electronic circuitry, which recently achieved the record setting length of 1 nanometer for a single gate.

SUBDISCIPLINES

Electrical engineering has many subdisciplines, the most common of which are listed below. Although there are electrical engineers who focus exclusively on one of these subdisciplines, many deal with a combination of them. Sometimes certain fields, such as electronic engineering and computer engineering, are considered separate disciplines in their own right.

Power

Power pole

Power engineering deals with the generation, transmission, and distribution of electricity as well as the design of a range of related devices. These include transformers, electric generators, electric motors, high voltage engineering, and power electronics. In many regions of the world, governments maintain an electrical network called a power grid that connects a variety of generators together with users of their energy. Users purchase electrical energy from the grid, avoiding the costly exercise of having to generate their own. Power engineers may work on the design and maintenance of the power grid as well as the power systems that connect to it. Such systems are called on-grid power systems and may supply the grid with additional power, draw power from the grid, or do both. Power engineers may also work on systems that do not connect to the grid, called off-grid power systems, which in some cases are preferable to on-grid systems. The future includes Satellite controlled power systems, with feedback in real time to prevent power surges and prevent blackouts.

Control

Control systems play a critical role in spaceflight.

Control engineering focuses on the modeling of a diverse range of dynamic systems and the design of controllers that will cause these systems to behave in the desired manner. To implement such controllers, electrical engineers may use electronic circuits, digital signal processors, microcontrollers, and programmable logic controls (PLCs). Control engineering has a wide range of applications from the flight and propulsion systems of commercial airliners to the cruise control present in many modern automobiles. It also plays an important role in industrial automation.

Control engineers often utilize feedback when designing control systems. For example, in an automobile with cruise control the vehicle's speed is continuously monitored and fed back to the system which adjusts the motor›s power output accordingly. Where there is regular feedback, control theory can be used to determine how the system responds to such feedback.

Electronics

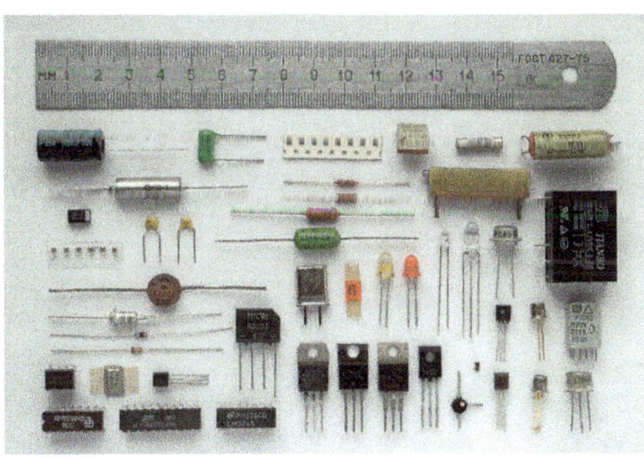

Electronic components

Electronic engineering involves the design and testing of electronic circuits that use the properties of components such as resistors, capacitors, inductors, diodes, and transistors to achieve a particular functionality. The tuned circuit, which allows the user of a radio to filter out all but a single station, is just one example of such a circuit. Another example (of a pneumatic signal conditioner) is shown in the adjacent photograph.

Prior to the Second World War, the subject was commonly known as radio engineering and basically was restricted to aspects of communications and radar, commercial radio, and early television. Later, in post war years, as consumer devices began to be developed, the field grew to include modern television, audio systems, computers, and microprocessors. In the mid-to-late 1950s, the term radio engineering gradually gave way to the name electronic engineering.

Before the invention of the integrated circuit in 1958, electronic circuits were constructed from discrete components that could be manipulated by humans. These discrete circuits consumed much space and power and were limited in speed, although they are still common in some applications. By contrast, integrated circuits packed a large number—often millions—of tiny electrical components, mainly transistors, into a small chip around the size of a coin. This allowed for the powerful computers and other electronic devices we see today.

Microelectronics

Microprocessor

Microelectronics engineering deals with the design and microfabrication of very small electronic circuit components for use in an integrated circuit or sometimes for use on their own as a general electronic component. The most common microelectronic components are semiconductor transistors, although all main electronic components (resistors, capacitors etc.) can be created at a microscopic level. Nanoelectronics is the further scaling of devices down to nanometer levels. Modern devices are already in the nanometer regime, with below 100 nm processing having been standard since about 2002.

Microelectronic components are created by chemically fabricating wafers of semiconductors such as silicon (at higher frequencies, compound semiconductors like gallium arsenide and indium phosphide) to obtain the desired transport of electronic charge and control of current. The field of microelectronics involves a significant amount of chemistry and material science and requires the electronic engineer working in the field to have a very good working knowledge of the effects of quantum mechanics.

Signal Processing

A Bayer filter on a CCD requires signal processing to get a red, green, and blue value at each pixel.

Signal processing deals with the analysis and manipulation of signals. Signals can be either analog, in which case the signal varies continuously according to the information, or digital, in which case the signal varies according to a series of discrete values representing the information. For analog signals, signal processing may involve the amplification and filtering of audio signals for audio equipment or the modulation and demodulation of signals for telecommunications. For digital signals, signal processing may involve the compression, error detection and error correction of digitally sampled signals.

Signal Processing is a very mathematically oriented and intensive area forming the core of digital signal processing and it is rapidly expanding with new applications in every field of electrical engineering such as communications, control, radar, audio engineering, broadcast engineering, power electronics, and biomedical engineering as many already existing analog systems are replaced with their digital counterparts. Analog signal processing is still important in the design of many control systems.

DSP processor ICs are found in every type of modern electronic systems and products including, SDTV | HDTV sets, radios and mobile communication devices, Hi-Fi audio equipment, Dolby noise reduction algorithms, GSM mobile phones, mp3 multimedia players, camcorders and digital cameras, automobile control systems, noise cancelling headphones, digital spectrum analyzers, intelligent missile guidance, radar, GPS based cruise control systems, and all kinds of image processing, video processing, audio processing, and speech processing systems.

Telecommunications

Satellite dishes are a crucial component in the analysis of satellite information.

Telecommunications engineering focuses on the transmission of information across a channel such as a coax cable, optical fiber or free space. Transmissions across free space require information to be encoded in a carrier signal to shift the information to a carrier frequency suitable for transmission; this is known as modulation. Popular analog modulation techniques include amplitude modulation and frequency modulation. The choice of modulation affects the cost and performance of a system and these two factors must be balanced carefully by the engineer.

Once the transmission characteristics of a system are determined, telecommunication engineers design the transmitters and receivers needed for such systems. These two are sometimes combined to form a two-way communication device known as a transceiver. A key consideration in the design of transmitters is their power consumption as this is closely related to their signal strength. If the signal strength of a transmitter is insufficient the signal›s information will be corrupted by noise.

Instrumentation

Flight instruments provide pilots with the tools to control aircraft analytically.

Instrumentation engineering deals with the design of devices to measure physical quantities such as pressure, flow, and temperature. The design of such instrumentation requires a good understanding of physics that often extends beyond electromagnetic theory. For example, flight instruments measure variables such as wind speed and altitude to enable pilots the control of aircraft analytically. Similarly, thermocouples use the Peltier-Seebeck effect to measure the temperature difference between two points.

Often instrumentation is not used by itself, but instead as the sensors of larger electrical systems. For example, a thermocouple might be used to help ensure a furnace's temperature remains constant. For this reason, instrumentation engineering is often viewed as the counterpart of control engineering.

Computers

Supercomputers are used in fields as diverse as computational biology and geographic information systems.

Computer engineering deals with the design of computers and computer systems. This may involve the design of new hardware, the design of PDAs, tablets, and supercomputers, or the use of computers to control an industrial plant. Computer engineers may also work on a system›s software. However, the design of complex software systems is often the domain of software engineering, which is usually considered a separate discipline. Desktop computers represent a tiny fraction of the devices a computer engineer might work on, as computer-like architectures are now found in a range of devices including video game consoles and DVD players.

Related Disciplines

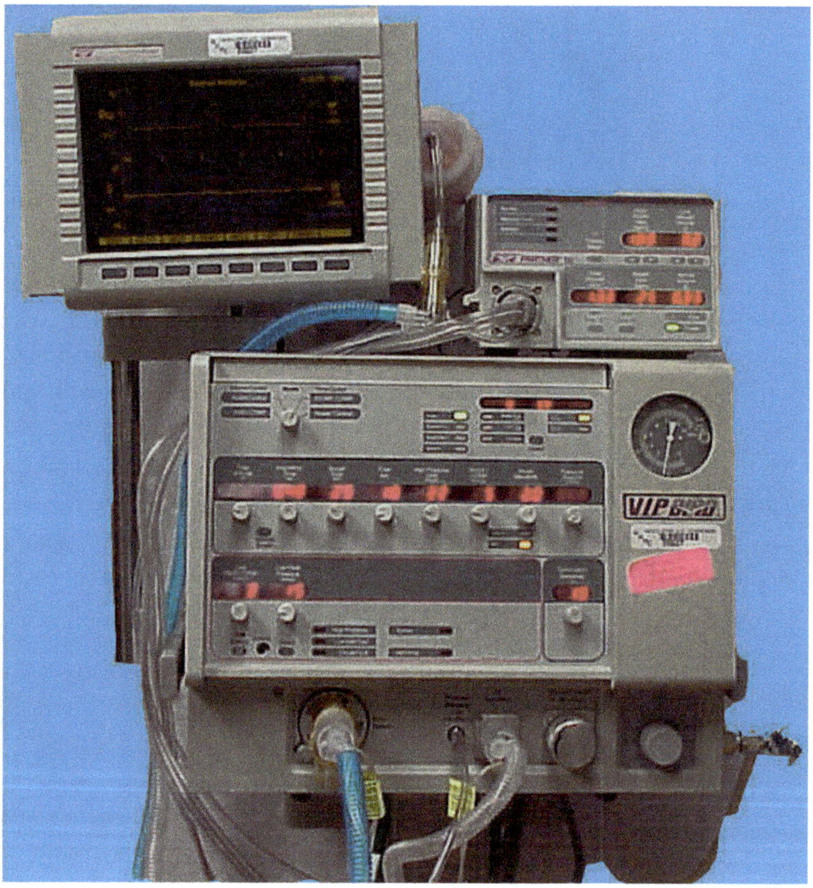

The Bird VIP Infant ventilator

Mechatronics is an engineering discipline which deals with the convergence of electrical and mechanical systems. Such combined systems are known as electromechanical systems and have widespread adoption. Examples include automated manufacturing systems, heating, ventilation and air-conditioning systems, and various subsystems of aircraft and automobiles.

The term mechatronics is typically used to refer to macroscopic systems but futurists have predicted the emergence of very small electromechanical devices. Already, such small devices, known as Microelectromechanical systems (MEMS), are used in automobiles to tell airbags when to deploy, in digital projectors to create sharper images, and in inkjet printers to create nozzles for high definition printing. In the future it is hoped the devices will help build tiny implantable medical devices and improve optical communication.

Biomedical engineering is another related discipline, concerned with the design of medical equipment. This includes fixed equipment such as ventilators, MRI scanners, and electrocardiograph monitors as well as mobile equipment such as cochlear implants, artificial pacemakers, and artificial hearts.

Aerospace engineering and robotics an example is the most recent electric propulsion and ion propulsion.

Education

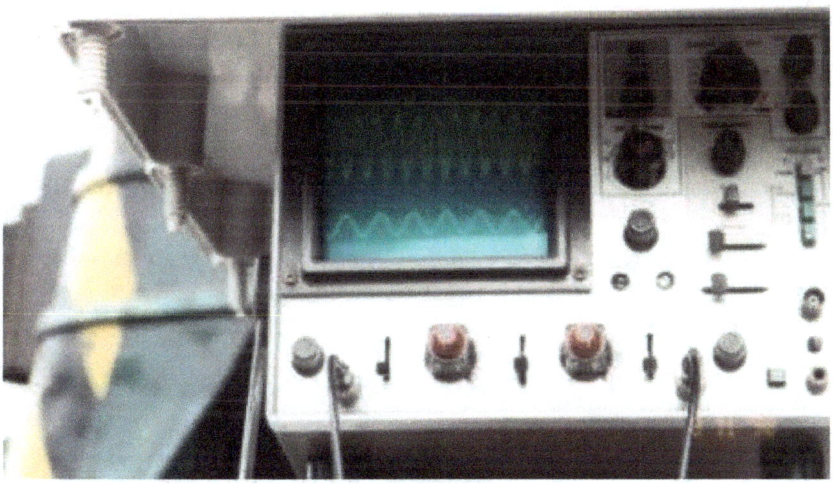

Oscilloscope

Electrical engineers typically possess an academic degree with a major in electrical engineering, electronics engineering, electrical engineering technology, or electrical and electronic engineering. The same fundamental principles are taught in all programs, though emphasis may vary according to title. The length of study for such a degree is usually four or five years and the completed degree may be designated as a Bachelor of Science in Electrical/Electronics Engineering Technology, Bachelor of Engineering, Bachelor of Science, Bachelor of Technology, or Bachelor of Applied Science depending on the university. The bachelor›s degree generally includes units covering physics, mathematics, computer science, project management, and a variety of topics in electrical engineering. Initially such topics cover most, if not all, of the subdisciplines of electrical engineering. At some schools, the students can then choose to emphasize one or more subdisciplines towards the end of their courses of study.

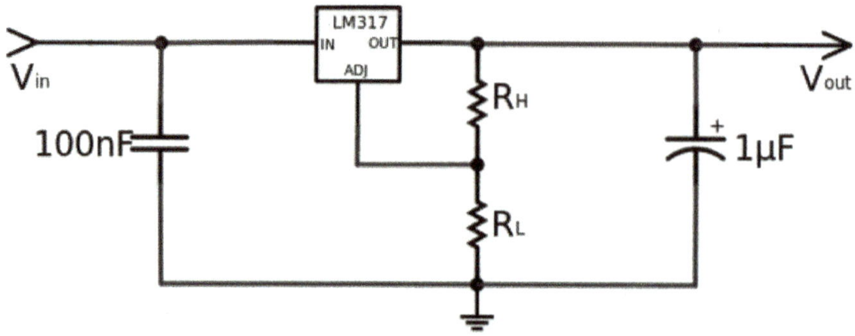

Typical electrical engineering diagram used as a troubleshooting tool

Practicing Engineers

Belgian electrical engineers inspecting the rotor of a 40,000 kilowatt turbine of the General Electric Company in New York City

In most countries, a bachelor›s degree in engineering represents the first step towards professional certification and the degree program itself is certified by a professional body. After completing a certified degree program the engineer must satisfy a range of requirements (including work experience requirements) before being certified. Once certified the engineer is designated the title of Professional Engineer (in the United States, Canada and South Africa), Chartered Engineer or Incorporated Engineer (in India, Pakistan, the United Kingdom, Ireland and Zimbabwe), Chartered Professional Engineer (in Australia and New Zealand) or European Engineer (in much of the European Union).

The IEEE corporate office is on the 17th floor of 3 Park Avenue in New York City

The advantages of certification vary depending upon location. For example, in the United States and Canada «only a licensed engineer may seal engineering work for public and private clients». This requirement is enforced by state and provincial legislation such as Quebec›s Engineers Act. In other countries, no such legislation exists. Practically all certifying bodies maintain a code of ethics that they expect all members to abide by or risk expulsion. In this way these organizations play an important role in maintaining ethical standards for the profession. Even in jurisdictions where certification has little or no legal bearing on work, engineers are subject to contract law. In cases where an engineer›s work fails he or she may be subject to the tort of negligence and, in extreme cases, the charge of criminal negligence. An engineer›s work must also comply with numerous other rules and regulations such as building codes and legislation pertaining to environmental law.

Professional bodies of note for electrical engineers include the Institute of Electrical and Electronics Engineers (IEEE) and the Institution of Engineering and Technology (IET). The IEEE claims to produce 30% of the world›s literature in electrical engineering, has over 360,000 members worldwide and holds over 3,000 conferences annually. The IET publishes 21 journals, has a worldwide membership of over 150,000, and claims to be the largest professional engineering society in Europe. Obsolescence of technical skills is a serious concern for electrical engineers. Membership and participation in technical societies, regular reviews of periodicals in the field and a habit of continued learning are therefore essential to maintaining proficiency. An MIET(Member of the Institution of Engineering and Technology) is recognised in Europe as an Electrical and computer (technology) engineer.

In Australia, Canada, and the United States electrical engineers make up around 0.25% of the labor force (see note).

Tools and Work

From the Global Positioning System to electric power generation, electrical engineers have contributed to the development of a wide range of technologies. They design, develop, test, and supervise the deployment of electrical systems and electronic devices. For example, they may work on the design of telecommunication systems, the operation of electric power stations, the lighting and wiring of buildings, the design of household appliances, or the electrical control of industrial machinery.

Satellite communications is typical of what electrical engineers work on.

Fundamental to the discipline are the sciences of physics and mathematics as these help to obtain both a qualitative and quantitative description of how such systems will work. Today most engineering work involves the use of computers and it is commonplace to use computer-aided design programs when designing electrical systems. Nevertheless, the ability to sketch ideas is still invaluable for quickly communicating with others.

The Shadow robot hand system

Although most electrical engineers will understand basic circuit theory (that is the interactions of elements such as resistors, capacitors, diodes, transistors, and inductors in a circuit), the theories employed by engineers generally depend upon the work they do. For example, quantum mechanics and solid state physics might be relevant to an engineer working on VLSI (the design of integrated circuits), but are largely irrelevant to engineers working with macroscopic electrical systems. Even circuit theory may not be relevant to a person designing telecommunication systems that use off-the-shelf components. Perhaps the most important technical skills for electrical engineers are reflected in university programs, which emphasize strong numerical skills, computer literacy, and the ability to understand the technical language and concepts that relate to electrical engineering.

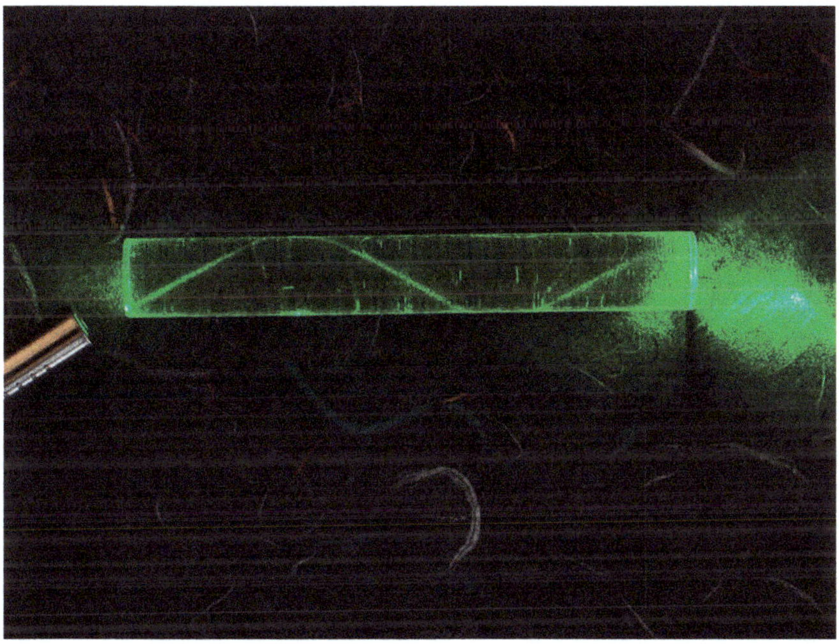

A laser bouncing down an acrylic rod, illustrating the total internal reflection of light in a multi-mode optical fiber.

A wide range of instrumentation is used by electrical engineers. For simple control circuits and alarms, a basic multimeter measuring voltage, current, and resistance may suffice. Where time-varying signals need to be studied, the oscilloscope is also an ubiquitous instrument. In RF engineering and high frequency telecommunications, spectrum analyzers and network analyzers are used. In some disciplines, safety can be a particular concern with instrumentation. For instance, medical electronics designers must take into account that much lower voltages than normal can be dangerous when electrodes are directly in contact with internal body fluids. Power transmission engineering also has great safety concerns due to the high voltages used; although voltmeters may in principle be similar to their low voltage equivalents, safety and calibration issues make them very different. Many disciplines of electrical engineering use tests specific to their discipline. Audio electronics engineers use audio test sets consisting of a signal generator and a meter, principally to measure level but also other parameters such as harmonic distortion and noise. Likewise, information technology have their own test sets, often specific to a particular data format, and the same is true of television broadcasting.

Radome at the Misawa Air Base Misawa Security Operations Center, Misawa, Japan

For many engineers, technical work accounts for only a fraction of the work they do. A lot of time may also be spent on tasks such as discussing proposals with clients, preparing budgets and determining project schedules.Many senior engineers manage a team of technicians or other engineers and for this reason project management skills are important. Most engineering projects involve some form of documentation and strong written communication skills are therefore very important.

The workplaces of engineers are just as varied as the types of work they do. Electrical engineers may be found in the pristine lab environment of a fabrication plant, the offices of a consulting firm or on site at a mine. During their working life, electrical engineers may find themselves supervising a wide range of individuals including scientists, electricians, computer programmers, and other engineers.

Electrical engineering has an intimate relationship with the physical sciences. For instance, the physicist Lord Kelvin played a major role in the engineering of the first transatlantic telegraph cable. Conversely, the engineer Oliver Heaviside produced major work on the mathematics of transmission on telegraph cables. Electrical engineers are often required on major science projects. For instance, large particle accelerators such as CERN need electrical engineers to deal with many aspects of the project: from the power distribution, to the instrumentation, to the manufacture and installation of the superconducting electromagnets.

ANCIENT DEVELOPMENTS

Thales of Miletus, an ancient Greek philosopher, writing at around 600 B.C.E., described a form of static electricity, noting that rubbing fur on various substances, such as amber, would cause a particular attraction between the two. He noted that the amber buttons could attract light objects such as hair and that if they rubbed the amber for long enough they could even get a spark to jump.

At around 450 B.C.E. Democritus, a later Greek philosopher, developed an atomic theory that was remarkably similar to our modern atomic theory. His mentor, Leucippus, is credited with this same theory. The hypothesis of Leucippus and Democritus held everything to be composed of atoms. But these atoms, called "atomos", were indivisible, and indestructible. He presciently stated that between atoms lies empty space, and that atoms are constantly in motion. He was incorrect only in stating that atoms come in different sizes and shapes. Each object had its own shaped and sized atom.

An object found in Iraq in 1938, dated to about 250 B.C.E. and called the Baghdad Battery, resembles a galvanic cell and is believed by some to have been used for electroplating in Mesopotamia, although this has not yet been proven.

17Th Century Developments

A voltaic pile, the first battery

Alessandro Volta showing the earliest pile to emperor Napoleon Bonaparte

Electricity would remain little more than an intellectual curiosity for millennia. In 1600, the English scientist, William Gilbert extended the study of Cardano on electricity and magnetism, distinguishing the lodestone effect from static electricity produced by rubbing amber. He coined the New Latin word electricus ("of amber" or "like amber", from ήλεκτρον [elektron], the Greek word for "amber") to refer to the property of attracting small objects after being rubbed.This association gave rise to the English words "electric" and "electricity", which made their first appearance in print in Thomas Browne's Pseudodoxia Epidemica of 1646.

Further work was conducted by Otto von Guericke who showed electrostatic repulsion. Robert Boyle also published work.

18Th Century Developments

By 1705, Francis Hauksbee had discovered that if he placed a small amount of mercury in the glass of his modified version of Otto von Guericke's generator, evacuated the air from it to create a mild vacuum and rubbed the ball in order to build up a charge, a glow was visible if he placed his hand on the outside of the ball. This glow was bright enough to read by. It seemed to be similar to St. Elmo's Fire. This effect later became the basis of the gas-discharge lamp, which led to neon lighting and mercury vapor lamps. In 1706 he produced an ‹Influence machine› to generate this effect. He was elected a Fellow of the Royal Society the same year.

Hauksbee continued to experiment with electricity, making numerous observations and developing machines to generate and demonstrate various electrical phenomena. In 1709 he published Physico-Mechanical Experiments on Various Subjects which summarized much of his scientific work.

Stephen Gray discovered the importance of insulators and conductors. C. F. du Fay seeing his work, developed a «two-fluid» theory of electricity.

Benjamin Franklin

In the 18th century, Benjamin Franklin conducted extensive research in electricity, selling his possessions to fund his work. In June 1752 he is reputed to have attached a metal key to the bottom of a dampened kite string and flown the kite in a storm-threatened sky. A succession of sparks jumping from the key to the back of his hand showed that lightning was indeed electrical in nature. He also explained the apparently paradoxical behavior of the Leyden jar as a device for storing large amounts of electrical charge, by coming up with the single fluid, two states theory of electricity.

In 1791, Italian Luigi Galvani published his discovery of bioelectricity, demonstrating that electricity was the medium by which nerve cells passed signals to the muscles. Alessandro Volta's battery, or voltaic pile, of 1800, made from alternating layers of zinc and copper, provided scientists with a more reliable source of electrical energy than the electrostatic machines previously used.

19Th Century Developments

Electrical engineering became a profession in the late 19th century. Practitioners had created a global electric telegraph network and the first electrical engineering institutions to support the new discipline were founded in the UK and USA. Although it is impossible to precisely pinpoint a first electrical engineer, Francis Ronalds stands ahead of the field, who created the first working electric telegraph system in 1816 and documented his vision of how the world could be transformed by electricity. Over 50 years later, he joined the new Society of Telegraph Engineers (soon to be renamed the Institution of Electrical Engineers) where he was regarded by other members as the first of their cohort. The donation of his extensive electrical library was a considerable boon for the fledgling Society.

Michael Faraday portrayed by Thomas Phillips c. 1841–1842

Development of the scientific basis for electrical engineering, with the tools of modern research techniques, intensified during the 19th century. Notable developments early in this century include the work of Georg Ohm, who in 1827 quantified the relationship between the electric current and potential difference in a conductor, Michael Faraday, the discoverer of electromagnetic induction in 1831. In the 1830s, Georg Ohm also constructed an early electrostatic machine. The homopolar generator was developed first by Michael Faraday during his memorable experiments in 1831. It was the beginning of modern dynamos — that is, electrical generators which operate using a magnetic field. The invention of the industrial generator, which didn't need external magnetic power in 1866 by Werner von Siemens made a large series of other inventions in the wake possible.

In 1873 James Clerk Maxwell published a unified treatment of electricity and magnetism in A Treatise on Electricity and Magnetism which stimulated several theorists to think in terms of fields described by Maxwell's equations. In 1878, the British inventor James Wimshurst developed an apparatus that had two glass disks mounted on two shafts. It was not till 1883 that the Wimshurst machine was more fully reported to the scientific community.

Thomas Edison built the world's first large-scale electrical supply network

During the latter part of the 1800s, the study of electricity was largely considered to be a subfield of physics. It was not until the late 19th century that universities started to offer degrees in electrical engineering. In 1882, Darmstadt University of Technology founded the first chair and the first faculty of electrical engineering worldwide. In the same year, under Professor Charles Cross, the Massachusetts Institute of Technology began offering the first option of Electrical Engineering within a physics department. In 1883, Darmstadt University of Technology and Cornell University introduced the world's first courses of study in electrical engineering and in 1885 the University College London founded the first chair of electrical engineering in the United Kingdom. The University of Missouri subsequently established the first department of electrical engineering in the United States in 1886.

During this period commercial use of electricity increased dramatically. Starting in the late 1870s cities started installing large scale electric street lighting systems based on arc lamps. After the development of a practical incandescent lamp for indoor lighting, Thomas Edison switched on the world's first public electric supply utility in 1882, using what was considered a relatively safe 110 volts direct current system to supply customers. Engineering advances in the 1880s, including the invention of the transformer, led to electric utilities starting to adopting alternating current, up till then used primarily in arc lighting systems, as a distribution standard for outdoor and indoor lighting (eventually replacing direct current for such purposes). In the US there was a rivalry, primarily between a Westinghouse AC and the Edison DC system known as the "War of Currents".

George Westinghouse, American entrepreneur and engineer, financially backed the development of a practical AC power network.

"By the mid-1890s the four "Maxwell equations" were recognized as the foundation of one of the strongest and most successful theories in all of physics; they had taken their place as companions, even rivals, to Newton's laws of mechanics. The equations were by then also being put to practical use, most dramatically in the emerging new technology of radio communications, but also in the telegraph, telephone, and electric power industries." By the end of the 19th century, figures in the progress of electrical engineering were beginning to emerge.

Charles Proteus Steinmetz helped foster the development of alternating current that made possible the expansion of the electric power industry in the United States, formulating mathematical theories for engineers.

Emergence of Radio and Electronics

Charles Proteus Steinmetz circa 1915

During the development of radio, many scientists and inventors contributed to radio technology and electronics. In his classic UHF experiments of 1888, Heinrich Hertz demonstrated the existence of electromagnetic waves (radio waves) leading many inventors and scientists to try to adapt them to commercial applications, such as Guglielmo Marconi (1895) and Alexander Popov (1896).

20Th Century Developments

John Fleming invented the first radio tube, the diode, in 1904.

Reginald Fessenden recognized that a continuous wave needed to be generated to make speech transmission possible, and by the end of 1906 he sent the first radio broadcast of voice. Also in 1906, Robert von Lieben and Lee De Forest independently developed the amplifier tube, called the triode. Edwin Howard Armstrong enabling technology for electronic television, in 1931.

Second World War Years

The second world war saw tremendous advances in the field of electronics; especially in radar and with the invention of the magnetron by Randall and Boot at the University of Birmingham in 1940. Radio location, radio communication and radio guidance of aircraft were all developed at this time. An early electronic computing device, Colossus was built by Tommy Flowers of the GPO to decipher the coded messages of the German Lorenz cipher machine. Also developed at this time were advanced clandestine radio transmitters and receivers for use by secret agents.

An American invention at the time was a device to scramble the telephone calls between Winston Churchill and Franklin D. Roosevelt. This was called the Green Hornet system and worked by inserting noise into the signal. The noise was then extracted at the receiving end. This system was never broken by the Germans.

A great amount of work was undertaken in the United States as part of the War Training Program in the areas of radio direction finding, pulsed linear networks, frequency modulation, vacuum tube circuits, transmission line theory and fundamentals of electromagnetic engineering. These studies were published shortly after the war in what became known as the 'Radio Communication Series' published by McGraw-Hill in 1946.

In 1941 Konrad Zuse presented the Z3, the world's first fully functional and programmable computer.

Post war Developments

Prior to the second world war the subject was commonly known as 'radio engineering' and basically was restricted to aspects of communications and radar, commercial radio and early television. At this time, study of radio engineering at universities could only be undertaken as part of a physics degree.

Later, in post war years, as consumer devices began to be developed, the field broadened to include modern TV, audio systems, Hi-Fi and latterly computers and microprocessors. In 1946 the ENIAC (Electronic Numerical Integrator and Computer) of John Presper Eckert and John Mauchly followed, beginning the computing era. The arithmetic performance of these machines allowed engineers to develop completely new technologies and achieve new objectives, including the Apollo missions and the NASA moon landing.

The invention of the transistor in 1947 by William B. Shockley, John Bardeen and Walter Brattain opened the door for more compact devices and led to the development of the integrated circuit in 1958 by Jack Kilby and independently in 1959 by Robert Noyce. In the mid to late 1950s, the term radio engineering gradually gave way to the name electronics engineering, which then became a stand-alone university degree subject, usually taught alongside electrical engineering with which it had become associated due to some similarities. In 1968 Marcian Hoff invented the first microprocessor at Intel and thus ignited the development of the personal computer. The first realization of the microprocessor was the Intel 4004, a 4-bit processor developed in 1971, but only in 1973 did the Intel 8080, an 8-bit processor, make the building of the first personal computer, the Altair 8800, possible.

REFERENCES

1. Russell, Bertrand (1972). A History of Western Philosophy, Simon & Schuster. pp.64-65.
2. Barnes, Jonathan.(1987). Early Greek Philosophy, Penguin.
3. Stewart, Joseph (2001), Intermediate Electromagnetic Theory, World Scientific, p. 50, ISBN 981-02-4471-1

4. Baigrie, Brian (2006), Electricity and Magnetism: A Historical Perspective, Greenwood Press, pp. 7–8, ISBN 0-313-33358-0
5. Chalmers, Gordon (1937), "The Lodestone and the Understanding of Matter in Seventeenth Century England", Philosophy of Science, 4 (1): 75–95, doi:10.1086/286445
6. Burke, James (1978). Connections. London: Macmillan. p. 75. ISBN 0-333-24827-9.
7. "Library and Archive Catalogue". Retrieved 2012-03-09.
8. Srodes, James (2002), Franklin: The Essential Founding Father, Regnery Publishing, pp. 92–94, ISBN 0-89526-163-4 It is uncertain if Franklin personally carried out this experiment, but it is popularly attributed to him.
9. Uman, Martin (1987), All About Lightning (PDF), Dover Publications, ISBN 0-486-25237-X
10. Kirby, Richard S. (1990), Engineering in History, Courier Dover Publications, pp. 331–333, ISBN 0-486-26412-2
11. Ronalds, B.F. (2016). Sir Francis Ronalds: Father of the Electric Telegraph. London: Imperial College Press. ISBN 978-1-78326-917-4.
12. Ronalds, B.F. (Feb 2016). "The Bicentennial of Francis Ronalds's Electric Telegraph". Physics Today. doi:10.1063/PT.3.3079.
13. Ronalds, B.F. (July 2016). "Francis Ronalds (1788-1873): The First Electrical Engineer?". Proceedings of the IEEE. doi:10.1109/JPROC.2016.2571358.
14. National Portrait gallery NPG 269
15. ""Ohm, Georg Simon", "Faraday, Michael"". Encyclopædia Britannica (11 ed.). 1911.
16. Weber, Ernst; Frederik Nebeker (1994). The Evolution of Electrical Engineering: A Personal Perspective. IEEE Press. ISBN 0-7803-1066-7.
17. Ryder, John; Donald Fink (1984). Engineers and Electrons. IEEE Press. ISBN 0-87942-172-X.
18. Quentin R. Skrabec, The 100 Most Significant Events in American Business: An Encyclopedia, ABC-CLIO - 2012, page 86
19. Quentin R. Skrabec Jr, The 100 Most Significant Events in American Business - 2012, Page 85
20. Bruce J. Hunt (1991) The Maxwellians, page one
21. "History". National Fire Protection Association. Retrieved January 19, 2006. (published 1996 in the NFPA Journal)
22. "History of Amateur Radio". What is Amateur Radio?. Retrieved January 18, 2006.
23. "History of TV". Retrieved January 18, 2006.
24. "The Z3". Retrieved January 18, 2006.
25. "The ENIAC Museum Online". Retrieved 2006-01-18.

26. "Electronics Timeline". Greatest Engineering Achievements of the Twentieth Century. Retrieved January 18, 2006.
27. "Computing History (1971 - 1975)". Retrieved January 18, 2006.
28. http://newscenter.lbl.gov/2016/10/06/smallest-transistor-1-nm-gate/
29. Martinsen & Grimnes 2011, p. 411.
30. Lambourne 2010, p. 11.
31. Ronalds, B.F. (2016). Sir Francis Ronalds: Father of the Electric Telegraph. London: Imperial College Press. ISBN 978-1-78326-917-4.
32. Ronalds, B.F. (2016). "Sir Francis Ronalds and the Electric Telegraph". Int. J. for the History of Engineering & Technology. doi:10.1080/17581206.2015.1119481.
33. Ronalds, B.F. (July 2016). "Francis Ronalds (1788-1873): The First Electrical Engineer?". Proceedings of the IEEE. doi:10.1109/JPROC.2016.2571358.
34. Rosenberg 2008, p. 9.
35. Tunbridge 1992.
36. Wildes & Lindgren 1985, p. 19.
37. https://www.ece.cornell.edu/ece/about/history.cfm
38. https://www.engineering.cornell.edu/about/upload/Cornell-Engineering-history.pdf
39. http://president.cornell.edu/andrew-dickson-white/
40. The Electrical Engineer. 1911. p. 54.
41. http://engineering.missouri.edu/ece/about/department-history/
42. Heertje & Perlman 1990, p. 138.
43. I. Grattan-Guinness, History and Philosophy of the Mathematical Sciences - 2003, Page 1229
44. Jeff Suzuki, Mathematics in Historical Context - 2009, page 329
45. Severs & Leise 2011, p. 145.
46. Marconi's biography at Nobelprize.org retrieved 21 June 2008.
47. Abramson 1955, p. 22.
48. Huurdeman 2003, p. 226.
49. "Albert W. Hull (1880–1966)". IEEE History Center. Retrieved 22 January 2006.
50. "Who Invented Microwaves?". Retrieved 22 January 2006.
51. "Early Radar History". Peneley Radar Archives. Retrieved 22 January 2006.
52. Rojas, Raúl (2002). "The history of Konrad Zuse's early computing machines". In Rojas, Raúl; Hashagen, Ulf. The First Computers—History and Architectures History of Computing. MIT Press. p. 237. ISBN 0-262-68137-4.
53. Sale, Anthony E. (2002). "The Colossus of Bletchley Park". In Rojas, Raúl; Hashagen, Ulf. The First Computers—History and Architectures History of Computing. MIT Press. pp. 354–355. ISBN 0-262-68137-4.
54. "The ENIAC Museum Online". Retrieved 18 January 2006.

55. "Electronics Timeline". Greatest Engineering Achievements of the Twentieth Century. Retrieved 18 January 2006.
56. "Computing History (1971–1975)". Retrieved 18 January 2006.
57. Grigsby 2012.
58. Engineering: Issues, Challenges and Opportunities for Development. UNESCO. 2010. pp. 127–8. ISBN 978-92-3-104156-3.
59. Bissell 1996, p. 17.
60. McDavid & Echaore-McDavid 2009, p. 95.
61. Fairman 1998, p. 119.
62. Thompson 2006, p. 4.
63. Merhari 2009, p. 233.
64. Bhushan 1997, p. 581.
65. Mook 2008, p. 149.
66. Sullivan 2012.
67. Tuzlukov 2010, p. 20.
68. Manolakis & Ingle 2011, p. 17.
69. Bayoumi & Swartzlander 1994, p. 25.
70. Khanna 2009, p. 297.
71. Tobin 2007, p. 15.
72. Chandrasekhar 2006, p. 21.
73. Smith 2007, p. 19.
74. Zhang, Hu & Luo 2007, p. 448.
75. Grant & Bixley 2011, p. 159.
76. Fredlund, Rahardjo & Fredlund 2012, p. 346.
77. Manual on the Use of Thermocouples in Temperature Measurement. ASTM International. 1 January 1993. p. 154. ISBN 978-0-8031-1466-1.
78. Obaidat, Denko & Woungang 2011, p. 9.
79. Jalote 2006, p. 22.
80. Mahalik 2003, p. 569.
81. Leondes 2000, p. 199.
82. Shetty & Kolk 2010, p. 36.
83. Maluf & Williams 2004, p. 3.
84. Iga & Kokubun 2010, p. 137.
85. Dodds, Kumar & Veering 2014, p. 274.

86. "Electrical and Electronic Engineer". Occupational Outlook Handbook, 2012-13 Edition. Bureau of Labor Statistics, U.S. Department of Labor. Retrieved November 15, 2014.

87. Chaturvedi 1997, p. 253.

88. "What is the difference between electrical and electronic engineering?". FAQs - Studying Electrical Engineering. Retrieved 20 March 2012.

89. Computerworld. IDG Enterprise. 25 August 1986. p. 97.

90. "Electrical and Electronic Engineering". Retrieved 8 December 2011.

91. Various including graduate degree requirements at MIT, study guide at UWA, the curriculum at Queen's and unit tables at Aberdeen

92. Occupational Outlook Handbook, 2008–2009. U S Department of Labor, Jist Works. 1 March 2008. p. 148. ISBN 978-1-59357-513-7.

93. "Why Should You Get Licensed?". National Society of Professional Engineers. Archived from the original on 4 June 2005. Retrieved 11 July 2005.

94. "Engineers Act". Quebec Statutes and Regulations (CanLII). Retrieved 24 July 2005.

95. "Codes of Ethics and Conduct". Online Ethics Center. Retrieved 24 July 2005.

96. "About the IEEE". IEEE. Retrieved 11 July 2005.

97. "About the IET". The IET. Retrieved 11 July 2005.

98. "Journal and Magazines". The IET. Retrieved 11 July 2005.

99. "Electrical and Electronics Engineers, except Computer". Occupational Outlook Handbook. Archived from the original on 13 July 2005. Retrieved 16 July 2005. (see here regarding copyright)

100. "Electrical and Electronics Engineers, except Computer". Occupational Outlook Handbook. Archived from the original on 13 July 2005. Retrieved 16 July 2005. (see Archived index at the Wayback Machine.)

101. Taylor 2008, p. 241.

102. Leitgeb 2010.

103. Naidu & Kamaraju 2009, p. 210

104. Trevelyan, James; (2005). What Do Engineers Really Do?. University of Western Australia. (seminar with slides)

105. McDavid & Echaore-McDavid 2009, p. 87.

106. Huurdeman, pp. 95 96

107. Huurdeman, p.90

108. Schmidt, p.218

109. Martini, p.179

110. "Electrical Engineers". www.bls.gov. Retrieved 2015-11-30.

111. sector=Government, ; corporateName=Department of Economic Development, Jobs, Transport and Resources - State Government of Victoria;. "Electrical

Engineer Career Information for Migrants | Victoria, Australia". www.liveinvictoria.vic.gov.au. Retrieved 2015-11-30.

112. "Electrical Engineers". Bureau of Labor Statistics. Retrieved 13 March 2009. See also: "Work Experience of the Population in 2006". Bureau of Labor Statistics. Retrieved 20 June 2008. and "Electrical and Electronics Engineers". Australian Careers. Retrieved 13 March 2009. and "Electrical and Electronics Engineers". Canadian jobs service. Retrieved 13 March 2009.

CHAPTER 3

Electromagnetism

Chapter Contents

Electromagnetism is a branch of physics which deals with electricity and magnetism and the interaction between them. It was first discovered in the 19th century and has extensive application in today's world of physics.

INTRODUCTION

Electromagnetism is a branch of physics involving the study of the electromagnetic force, a type of physical interaction that occurs between electrically charged particles. The electromagnetic force usually exhibits electromagnetic fields such as electric fields, magnetic fields, and light and is one of the four fundamental interactions (commonly called forces) in nature. The other three fundamental interactions are the strong interaction, the weak interaction, and gravitation.

Lightning is an electrostatic discharge that travels between two charged regions.ῄ

The word electromagnetism is a compound form of two Greek terms, ἤλεκτρον, ēlektron, "amber", and μαγνῆτις λίθος magnētis lithos, which means "magnesian stone", a type of iron ore. Electromagnetic phenomena are defined in terms of the electromagnetic force, sometimes called the Lorentz force, which includes both electricity and magnetism as different manifestations of the same phenomenon.

The electromagnetic force plays a major role in determining the internal properties of most objects encountered in daily life. Ordinary matter takes its form as a result of intermolecular forces between individual atoms and molecules in matter, and is a manifestation of the electromagnetic force. Electrons are bound by the electromagnetic force to atomic nuclei, and their orbital shapes and their influence on nearby atoms with their electrons is described by quantum mechanics. The electromagnetic force governs the processes involved in chemistry, which arise from interactions between the electrons of neighboring atoms.

There are numerous mathematical descriptions of the electromagnetic field. In classical electrodynamics, electric fields are described as electric potential and electric current. In Faraday's law, magnetic fields are associated with electromagnetic induction and magnetism, and Maxwell's equations describe how electric and magnetic fields are generated and altered by each other and by charges and currents.

The theoretical implications of electromagnetism, in particular the establishment of the speed of light based on properties of the "medium" (permeability and permittivity), led to the development of special relativity by Albert Einstein in 1905.

Although electromagnetism is considered one of the four fundamental forces, at high energy the weak force and electromagnetic force are unified as a single electroweak force. In the history of the universe, during the quark epoch the unified force broke into the two separate forces as the universe cooled.

LORENTZ FORCE

In physics (particularly in electromagnetism) the **Lorentz force** is the combination of electric and magnetic force on a point charge due to electromagnetic fields. If a particle of charge q moves with velocity **v** in the presence of an electric field **E** and a magnetic field **B**, then it will experience a force

$$\mathbf{F} = q\mathbf{E} + q\mathbf{v} \times \mathbf{B}$$

(in SI units). Variations on this basic formula describe the magnetic force on a current-carrying wire (sometimes called Laplace force), the electromotive force in a wire loop moving through a magnetic field (an aspect of Faraday's law of induction), and the force on a charged particle which might be travelling near the speed of light (relativistic form of the Lorentz force).

The first derivation of the Lorentz force is commonly attributed to Oliver Heaviside in 1889, although other historians suggest an earlier origin in an 1865 paper by James Clerk Maxwell. Hendrik Lorentz derived it a few years after Heaviside.

Charged Particle

The force **F** acting on a particle of electric charge q with instantaneous velocity **v**, due to an external electric field **E** and magnetic field **B**, is given by:

$$\mathbf{F} = q(\mathbf{E} + \mathbf{v} \times \mathbf{B})$$

where × is the vector cross product. All boldface quantities are vectors. More explicitly stated:

$$\mathbf{F}(\mathbf{r}, \dot{\mathbf{r}}, t, q) = q[\mathbf{E}(\mathbf{r}, t) + \dot{\mathbf{r}} \times \mathbf{B}(\mathbf{r}, t)]$$

in which r is the position vector of the charged particle, t is time, and the overdot is a time derivative.

A positively charged particle will be accelerated in the same linear orientation as the E field, but will curve perpendicularly to both the instantaneous velocity vector v and the B field according to the right-hand rule (in detail, if the fingers of the right hand are extended to point in the direction of v and are then curled to point in the direction of B, then the extended thumb will point in the direction of F).

The term qE is called the electric force, while the term qv × B is called the magnetic force. According to some definitions, the term "Lorentz force" refers specifically to the formula for the magnetic force, with the total electromagnetic force (including the electric force) given some other (nonstandard) name. This article will not follow this nomenclature: In what follows, the term "Lorentz force" will refer only to the expression for the total force.

The magnetic force component of the Lorentz force manifests itself as the force that acts on a current-carrying wire in a magnetic field. In that context, it is also called the Laplace force.

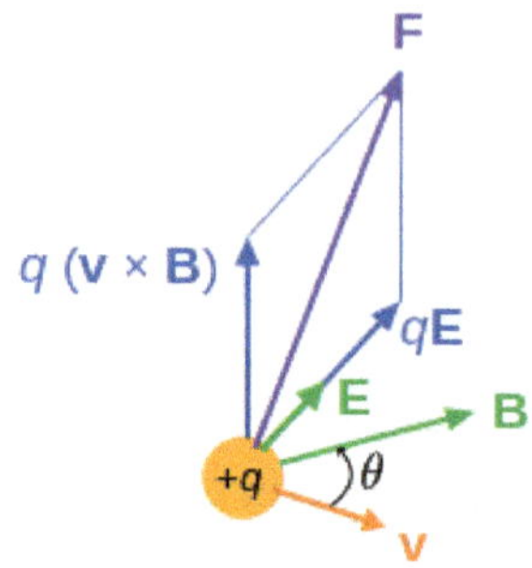

Lorentz force F on a charged particle (of charge q) in motion (instantaneous velocity v). The E field and B field vary in space and time.

Continuous Charge Distribution

For a continuous charge distribution in motion, the Lorentz force equation becomes:

$$\mathbf{dF} = \mathrm{d}q\,(\mathbf{E} + \mathbf{v} \times \mathbf{B})$$

where dF is the force on a small piece of the charge distribution with charge dq. If both sides of this equation are divided by the volume of this small piece of the charge distribution dV, the result is:

$$\mathbf{f} = \rho\,(\mathbf{E} + \mathbf{v} \times \mathbf{B})$$

where f is the force density (force per unit volume) and ρ is the charge density (charge per unit volume). Next, the current density corresponding to the motion of the charge continuum is

$$\mathbf{J} = \rho\mathbf{v}$$

so the continuous analogue to the equation is

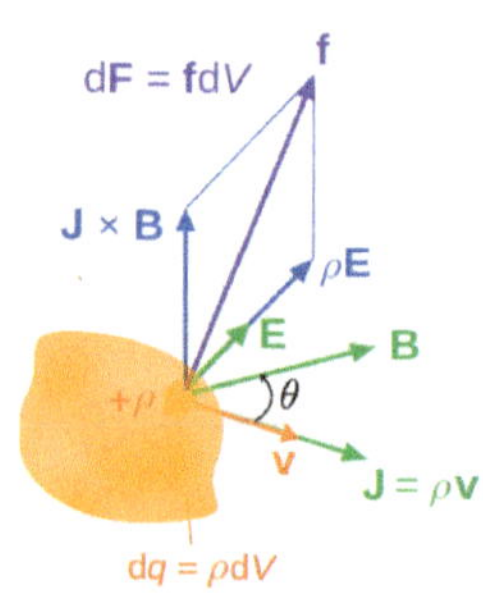

Lorentz force (per unit 3-volume) f on a continuous charge distribution (charge density ρ) in motion. The 3-current density J corresponds to the motion of the charge element dq in volume element dV and varies throughout the continuum.

$$\mathbf{f} = \rho\mathbf{E} + \mathbf{J} \times \mathbf{B}$$

The total force is the volume integral over the charge distribution:

$$\mathbf{F} = \iiint (\rho\mathbf{E} + \mathbf{J} \times \mathbf{B})\, \mathrm{d}V.$$

By eliminating ρ and J, using Maxwell's equations, and manipulating using the theorems of vector calculus, this form of the equation can be used to derive the Maxwell stress tensor σ, in turn this can be combined with the Poynting vector S to obtain the electromagnetic stress–energy tensor T used in general relativity.

In terms of σ and S, another way to write the Lorentz force (per unit volume) is

$$\mathbf{f} = \nabla \cdot \boldsymbol{\sigma} - \frac{1}{c^2}\frac{\partial \mathbf{S}}{\partial t}$$

where c is the speed of light and ∇· denotes the divergence of a tensor field. Rather than the amount of charge and its velocity in electric and magnetic fields, this equation relates the energy flux (flow of energy per unit time per unit distance) in the fields to the force exerted on a charge distribution. See Covariant formulation of classical electromagnetism for more details.

Early attempts to quantitatively describe the electromagnetic force were made in the mid-18th century. It was proposed that the force on magnetic poles, by Johann Tobias Mayer and others in 1760,and electrically charged objects, by Henry Cavendish in 1762, obeyed an inverse-square law. However, in both cases the experimental proof was neither complete nor conclusive. It was not until 1784 when Charles-Augustin de Coulomb, using a torsion balance, was able to definitively show through experiment that this was true. Soon after the discovery in 1820 by H. C. Ørsted that a magnetic needle is acted on by a voltaic current, André-Marie Ampère that same year was able to devise through experimentation the formula for the angular dependence of the force between two current elements. In all these descriptions, the force was always given in terms of the properties of the objects involved and the distances between them rather than in terms of electric and magnetic fields.

The modern concept of electric and magnetic fields first arose in the theories of Michael Faraday, particularly his idea of lines of force, later to be given full mathematical description by Lord Kelvin and James Clerk Maxwell. From a modern perspective it is possible to identify in Maxwell›s 1865 formulation of his field equations a form of the Lorentz force equation in relation to electric currents, however, in the time of Maxwell it was not evident how his equations related to the forces on moving charged objects. J. J. Thomson was the first to attempt to derive from Maxwell›s field equations the electromagnetic forces on a moving charged object in terms of the object›s properties and external fields. Interested in determining the electromagnetic behavior of the charged particles in cathode rays, Thomson published a paper in 1881 wherein he gave the force on the particles due to an external magnetic field as

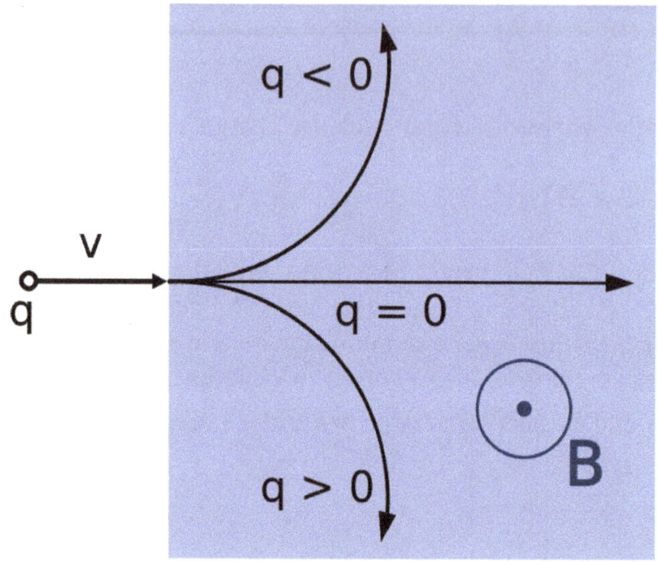

Trajectory of a particle with a positive or negative charge q under the influence of a magnetic field B, which is directed perpendicularly out of the screen.

Beam of electrons moving in a circle, due to the presence of a magnetic field. Purple light is emitted along the electron path, due to the electrons colliding with gas molecules in the bulb. A Teltron tube is used in this example.

$$\mathbf{F} = \frac{q}{2}\mathbf{v} \times \mathbf{B}.$$

Thomson derived the correct basic form of the formula, but, because of some miscalculations and an incomplete description of the displacement current, included an incorrect scale-factor of a half in front of the formula. Oliver Heaviside invented the modern vector notation and applied it to Maxwell's field equations; he also (in 1885 and 1889) had fixed the mistakes of Thomson's derivation and arrived at the correct form of the magnetic force on a moving charged object. Finally, in 1892, Hendrik Lorentz derived the modern form of the formula for the electromagnetic force which includes the contributions to the total force from both the electric and the magnetic fields. Lorentz began by abandoning the Maxwellian descriptions of the ether and conduction. Instead, Lorentz made a distinction between matter and the luminiferous aether and sought to apply the Maxwell equations at a microscopic scale. Using Heaviside's version of the Maxwell equations for a stationary ether and applying Lagrangian mechanics (see below), Lorentz arrived at the correct and complete form of the force law that now bears his name.

Trajectories of Particles Due to the Lorentz Force

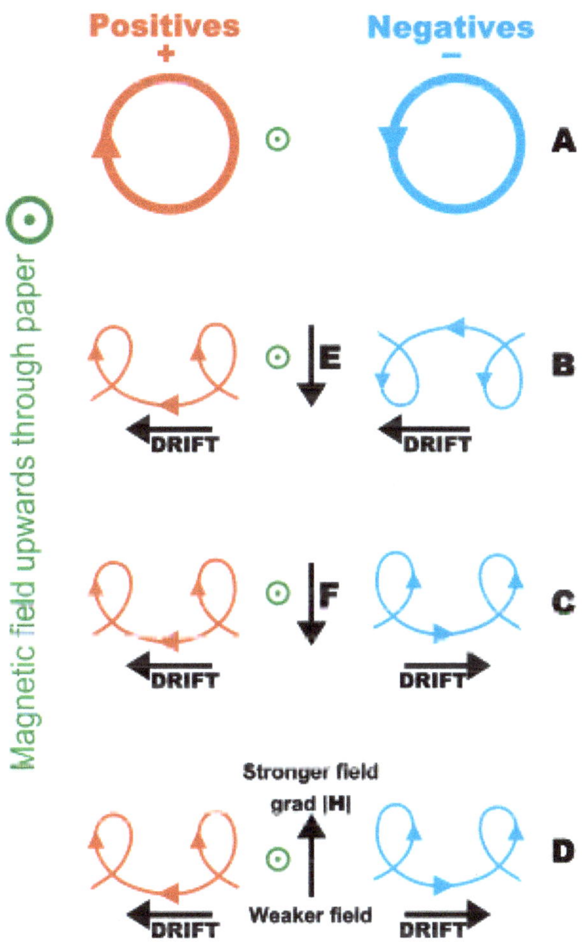

Charged particle drifts in a homogeneous magnetic field. (A) No disturbing force (B) With an electric field, E (C) With an independent force, F (e.g. gravity) (D) In an inhomogeneous magnetic field, grad H

In many cases of practical interest, the motion in a magnetic field of an electrically charged particle (such as an electron or ion in a plasma) can be treated as the superposition of a relatively fast circular motion around a point called the guiding center and a relatively slow drift of this point. The drift speeds may differ for various species depending on their charge states, masses, or temperatures, possibly resulting in electric currents or chemical separation.

Significance of the Lorentz Force

While the modern Maxwell's equations describe how electrically charged particles and currents or moving charged particles give rise to electric and magnetic fields, the Lorentz force law completes that picture by describing the force acting on a moving point charge q in the presence of electromagnetic fields. The Lorentz force law describes the effect of **E** and **B** upon a point charge, but such electromagnetic forces are not the entire picture. Charged particles are possibly coupled to other forces, notably gravity and nuclear forces. Thus, Maxwell's equations do not stand separate from other physical laws, but are coupled to them via the charge and current densities. The response of a point charge to the Lorentz law is one aspect; the generation of **E** and **B** by currents and charges is another.

In real materials the Lorentz force is inadequate to describe the collective behavior of charged particles, both in principle and as a matter of computation. The charged particles in a material medium not only respond to the **E** and **B** fields but also generate these fields. Complex transport equations must be solved to determine the time and spatial response of charges, for example, the Boltzmann equation or the Fokker–Planck equation or the Navier–Stokes equations. For example, see magnetohydrodynamics, fluid dynamics, electrohydrodynamics, superconductivity, stellar evolution. An entire physical apparatus for dealing with these matters has developed. See for example, Green–Kubo relations and Green›s function (many-body theory).

Lorentz Force Law as the Definition of E and B

In many textbook treatments of classical electromagnetism, the Lorentz force Law is used as the definition of the electric and magnetic fields **E** and **B**. To be specific, the Lorentz force is understood to be the following empirical statement:

The electromagnetic force **F** on a test charge at a given point and time is a certain function of its charge q and velocity **v**, which can be parameterized by exactly two vectors **E** and **B**, in the functional form:

$$\mathbf{F} = q(\mathbf{E} + \mathbf{v} \times \mathbf{B})$$

This is valid, even for particles approaching the speed of light (that is, magnitude of **v** = |**v**| = c). So the two vector fields **E** and **B** are thereby defined throughout space and time, and these are called the "electric field" and "magnetic field". The fields are defined everywhere in space and time with respect to what force a test charge would receive regardless of whether a charge is present to experience the force.

As a definition of **E** and **B**, the Lorentz force is only a definition in principle because a real particle (as opposed to the hypothetical "test charge" of infinitesimally-small mass and charge) would generate its own finite **E** and **B** fields, which would alter the electromagnetic force that it experiences. In addition, if the charge experiences acceleration, as if forced into a curved trajectory by some external agency, it emits radiation that causes braking of its motion. See for example Bremsstrahlung and synchrotron light. These effects occur through both a direct effect (called the radiation reaction force) and indirectly (by affecting the motion of nearby charges and currents). Moreover, net force must include gravity, electroweak, and any other forces aside from electromagnetic force.

Force on a Current-Carrying Wire

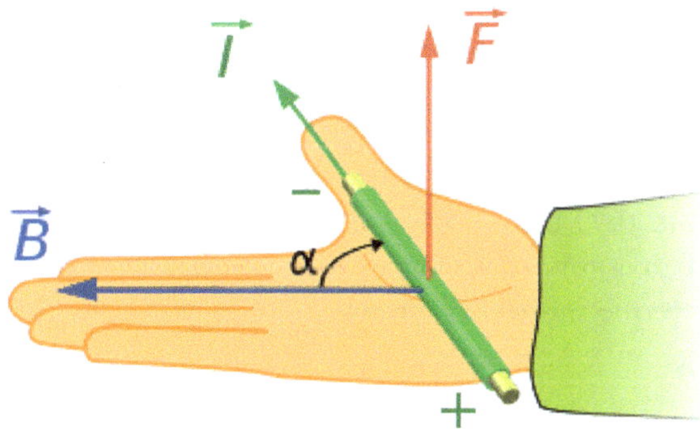

Right-hand rule for a current-carrying wire in a magnetic field B

When a wire carrying an electric current is placed in a magnetic field, each of the moving charges, which comprise the current, experiences the Lorentz force, and together they can create a macroscopic force on the wire (sometimes called the **Laplace force**). By combining the Lorentz force law above with the definition of electric current, the following equation results, in the case of a straight, stationary wire:

$$\mathbf{F} = I\boldsymbol{\ell} \times \mathbf{B}$$

where $\boldsymbol{\ell}$ is a vector whose magnitude is the length of wire, and whose direction is along the wire, aligned with the direction of conventional current flow I.

If the wire is not straight but curved, the force on it can be computed by applying this formula to each infinitesimal segment of wire $d\boldsymbol{\ell}$, then adding up all these forces by integration. Formally, the net force on a stationary, rigid wire carrying a steady current I is

$$\mathbf{F} = I \int d\boldsymbol{\ell} \times \mathbf{B}$$

This is the net force. In addition, there will usually be torque, plus other effects if the wire is not perfectly rigid.

One application of this is Ampère's force law, which describes how two current-carrying wires can attract or repel each other, since each experiences a Lorentz force from the other's magnetic field. For more information, see the article: Ampère's force law.

EMF

The magnetic force (q **v** × **B**) component of the Lorentz force is responsible for motional electromotive force (or motional EMF), the phenomenon underlying many electrical generators. When a conductor is moved through a magnetic field, the magnetic force tries to push electrons through the wire, and this creates the EMF. The term "motional EMF" is applied to this phenomenon, since the EMF is due to the motion of the wire.

In other electrical generators, the magnets move, while the conductors do not. In this case, the EMF is due to the electric force (q**E**) term in the Lorentz Force equation. The electric field in question is created by the changing magnetic field, resulting in an induced EMF, as described by the Maxwell–Faraday equation (one of the four modern Maxwell's equations).

Both of these EMF›s, despite their different origins, can be described by the same equation, namely, the EMF is the rate of change of magnetic flux through the wire. (This is Faraday›s law of induction, see below.) Einstein›s special theory of relativity was partially motivated by the desire to better understand this link between the two effects. In fact, the electric and magnetic fields are different faces of the same electromagnetic field, and in moving from one inertial frame to another, the solenoidal vector field portion of the E-field can change in whole or in part to a B-field or vice versa.

Lorentz Force and Faraday's Law of Induction

Lorentz force -image on a wall in Leiden

Given a loop of wire in a magnetic field, Faraday's law of induction states the induced electromotive force (EMF) in the wire is:

$$\mathcal{E} = -\frac{\mathrm{d}\Phi_B}{\mathrm{d}t}$$

where

$$\Phi_B = \iint_{\Sigma(t)} \mathrm{d}\mathbf{A} \cdot \mathbf{B}(\mathbf{r}, t)$$

is the magnetic flux through the loop, **B** is the magnetic field, Σ(t) is a surface bounded by the closed contour ∂Σ(t), at all at time t, d**A** is an infinitesimal vector area element of Σ(t) (magnitude is the area of an infinitesimal patch of surface, direction is orthogonal to that surface patch).

The sign of the EMF is determined by Lenz's law. Note that this is valid for not only a stationary wire – but also for a moving wire.

From Faraday›s law of induction (that is valid for a moving wire, for instance in a motor) and the Maxwell Equations, the Lorentz Force can be deduced. The reverse is also true, the Lorentz force and the Maxwell Equations can be used to derive the Faraday Law.

Let Σ(t) be the moving wire, moving together without rotation and with constant velocity **v** and Σ(t) be the internal surface of the wire. The EMF around the closed path ∂Σ(t) is given by:

$$\mathcal{E} = \oint_{\partial\Sigma(t)} d\boldsymbol{\ell} \cdot \mathbf{F}/q$$

where

$$\mathbf{E} = \mathbf{F}/q$$

is the electric field and dℓ is an infinitesimal vector element of the contour ∂Σ(t).

NB: Both dℓ and dA have a sign ambiguity; to get the correct sign, the right-hand rule is used, as explained in the article Kelvin–Stokes theorem.

The above result can be compared with the version of Faraday's law of induction that appears in the modern Maxwell's equations, called here the Maxwell–Faraday equation:

$$\nabla \times \mathbf{E} = -\frac{\partial \mathbf{B}}{\partial t}.$$

The Maxwell–Faraday equation also can be written in an integral form using the Kelvin–Stokes theorem.

So we have, the Maxwell Faraday equation:

$$\oint_{\partial\Sigma(t)} d\boldsymbol{\ell} \cdot \mathbf{E}(\mathbf{r},\, t) = -\iint_{\Sigma(t)} d\mathbf{A} \cdot \frac{d\mathbf{B}(\mathbf{r},\, t)}{dt}$$

and the Faraday Law,

$$\oint_{\partial\Sigma(t)} d\boldsymbol{\ell} \cdot \mathbf{F}/q(\mathbf{r},\, t) = -\frac{d}{dt}\iint_{\Sigma(t)} d\mathbf{A} \cdot \mathbf{B}(\mathbf{r},\, t).$$

The two are equivalent if the wire is not moving. Using the Leibniz integral rule and that div B = 0, results in,

$$\oint_{\partial\Sigma(t)} d\boldsymbol{\ell} \cdot \mathbf{F}/q(\mathbf{r},t) = -\iint_{\Sigma(t)} d\mathbf{A} \cdot \frac{\partial}{\partial t}\mathbf{B}(\mathbf{r},t) + \oint_{\partial\Sigma(t)} \mathbf{v} \times \mathbf{B}\, d\boldsymbol{\ell}$$

and using the Maxwell Faraday equation,

$$\oint_{\partial\Sigma(t)} d\boldsymbol{\ell} \cdot \mathbf{F}/q(\mathbf{r},\, t) = \oint_{\partial\Sigma(t)} d\boldsymbol{\ell} \cdot \mathbf{E}(\mathbf{r},\, t) + \oint_{\partial\Sigma(t)} \mathbf{v} \times \mathbf{B}(\mathbf{r},\, t)\, d\boldsymbol{\ell}$$

since this is valid for any wire position it implies that,

$$\mathbf{F} = q\mathbf{E}(\mathbf{r},\, t) + q\mathbf{v} \times \mathbf{B}(\mathbf{r},\, t).$$

Faraday's law of induction holds whether the loop of wire is rigid and stationary, or in motion or in process of deformation, and it holds whether the magnetic field is constant in time or changing. However, there are cases where Faraday's law is either inadequate or difficult to use, and application of the underlying Lorentz force law is necessary. See inapplicability of Faraday's law.

If the magnetic field is fixed in time and the conducting loop moves through the field, the magnetic flux ΦB linking the loop can change in several ways. For example, if the B-field varies with position, and the loop moves to a location with different B-field, ΦB will change. Alternatively, if the loop changes orientation with respect to the B-field, the B • dA differential element will change because of the different angle between B and dA, also changing ΦB. As a third example, if a portion of the circuit is swept through a uniform, time-independent B-field, and another portion of the circuit is held stationary, the flux linking the entire closed circuit can change due to the shift in relative position of the circuit's component parts with time (surface $\partial\Sigma(t)$ time-dependent). In all three cases, Faraday's law of induction then predicts the EMF generated by the change in ΦB.

Note that the Maxwell Faraday's equation implies that the Electric Field E is non conservative when the Magnetic Field B varies in time, and is not expressible as the gradient of a scalar field, and not subject to the gradient theorem since its rotational is not zero.

Lorentz Force in Terms Of Potentials

The E and B fields can be replaced by the magnetic vector potential A and (scalar) electrostatic potential ϕ by

$$\mathbf{E} = -\nabla\phi - \frac{\partial \mathbf{A}}{\partial t}$$

$$\mathbf{B} = \nabla \times \mathbf{A}$$

where ∇ is the gradient, ∇· is the divergence, ∇× is the curl.

The force becomes

$$\mathbf{F} = q\left[-\nabla\phi - \frac{\partial \mathbf{A}}{\partial t} + \mathbf{v} \times (\nabla \times \mathbf{A})\right]$$

and using an identity for the triple product simplifies to

$$\mathbf{F} = q\left[-\nabla\phi - \frac{\partial \mathbf{A}}{\partial t} + \nabla(\mathbf{v} \cdot \mathbf{A}) - (\mathbf{v} \cdot \nabla)\mathbf{A}\right]$$

using the chain rule, the total derivative of A is:

$$\frac{d\mathbf{A}}{dt} = \frac{\partial \mathbf{A}}{\partial t} + (\mathbf{v} \cdot \nabla)\mathbf{A}$$

so the above expression can be rewritten as;

$$\mathbf{F} = q\left[-\nabla(\phi - \mathbf{v} \cdot \mathbf{A}) - \frac{d\mathbf{A}}{dt}\right]$$

which can take the convenient Euler–Lagrange form

$$\mathbf{F} = q\left[-\nabla_{\mathbf{x}}(\phi - \dot{\mathbf{x}} \cdot \mathbf{A}) + \frac{d}{dt}\nabla_{\dot{\mathbf{x}}}(\phi - \dot{\mathbf{x}} \cdot \mathbf{A})\right]$$

Lorentz Force and Analytical Mechanics

The Lagrangian for a charged particle of mass m and charge q in an electromagnetic field equivalently describes the dynamics of the particle in terms of its energy, rather than the force exerted on it. The classical expression is given by:

$$L = \frac{m}{2}\dot{\mathbf{r}} \cdot \dot{\mathbf{r}} + q\mathbf{A} \cdot \dot{\mathbf{r}} - q\phi$$

where A and ϕ are the potential fields as above. Using Lagrange's equations, the equation for the Lorentz force can be obtained.

Derivation of Lorentz force from classical Lagrangian (SI units)

For an A field, a particle moving with velocity $\mathbf{v} = \dot{\mathbf{r}}$ has potential momentum $q\mathbf{A}(\mathbf{r}, t)$, so its potential energy is $q\mathbf{A}(\mathbf{r}, t) \cdot \dot{\mathbf{r}}$. For a ϕ field, the particle's potential energy is $q\phi(\mathbf{r}, t)$. The total potential energy is then:

$$V = q\phi - q\mathbf{A} \cdot \dot{\mathbf{r}}$$

and the kinetic energy is:

$$T = \frac{m}{2}\dot{\mathbf{r}} \cdot \dot{\mathbf{r}}$$

hence the Lagrangian:

$$L = T - V = \frac{m}{2}\dot{\mathbf{r}} \cdot \dot{\mathbf{r}} + q\mathbf{A} \cdot \dot{\mathbf{r}} - q\phi$$

$$L = \frac{m}{2}(\dot{x}^2 + \dot{y}^2 + \dot{z}^2) + q(\dot{x}A_x + \dot{y}A_y + \dot{z}A_z) - q\phi$$

Lagrange's equations are

$$\frac{d}{dt}\frac{\partial L}{\partial \dot{x}} = \frac{\partial L}{\partial x}$$

(same for y and z). So calculating the partial derivatives:

$$\begin{aligned}\frac{d}{dt}\frac{\partial L}{\partial \dot{x}} &= m\ddot{x} + q\frac{dA_x}{dt} \\ &= m\ddot{x} + \frac{q}{dt}\left(\frac{\partial A_x}{\partial t}dt + \frac{\partial A_x}{\partial x}dx + \frac{\partial A_x}{\partial y}dy + \frac{\partial A_x}{\partial z}dz\right) \\ &= m\ddot{x} + q\left(\frac{\partial A_x}{\partial t} + \frac{\partial A_x}{\partial x}\dot{x} + \frac{\partial A_x}{\partial y}\dot{y} + \frac{\partial A_x}{\partial z}\dot{z}\right)\end{aligned}$$

$$\frac{\partial L}{\partial x} = -q\frac{\partial \phi}{\partial x} + q\left(\frac{\partial A_x}{\partial x}\dot{x} + \frac{\partial A_y}{\partial x}\dot{y} + \frac{\partial A_z}{\partial x}\dot{z}\right)$$

equating and simplifying:

$$m\ddot{x} + q\left(\frac{\partial A_x}{\partial t} + \frac{\partial A_x}{\partial x}\dot{x} + \frac{\partial A_x}{\partial y}\dot{y} + \frac{\partial A_x}{\partial z}\dot{z}\right) = -q\frac{\partial \phi}{\partial x} + q\left(\frac{\partial A_x}{\partial x}\dot{x} + \frac{\partial A_y}{\partial x}\dot{y} + \frac{\partial A_z}{\partial x}\dot{z}\right)$$

$$\begin{aligned} F_x &= -q\left(\frac{\partial \phi}{\partial x} + \frac{\partial A_x}{\partial t}\right) + q\left[\dot{y}\left(\frac{\partial A_y}{\partial x} - \frac{\partial A_x}{\partial y}\right) + \dot{z}\left(\frac{\partial A_z}{\partial x} - \frac{\partial A_x}{\partial z}\right)\right] \\ &= qE_x + q[\dot{y}(\nabla \times \mathbf{A})_z - \dot{z}(\nabla \times \mathbf{A})_y] \\ &= qE_x + q[\dot{\mathbf{r}} \times (\nabla \times \mathbf{A})]_x \\ &= qE_x + q(\dot{\mathbf{r}} \times \mathbf{B})_x \end{aligned}$$

and similarly for the *y* and *z* directions. Hence the force equation is:

$$\mathbf{F} = q(\mathbf{E} + \dot{\mathbf{r}} \times \mathbf{B})$$

The potential energy depends on the velocity of the particle, so the force is velocity dependent, so it is not conservative.

The relativistic Lagrangian is

$$L = -mc^2\sqrt{1 - \left(\frac{\dot{\mathbf{r}}}{c}\right)^2} + e\mathbf{A}(\mathbf{r}) \cdot \dot{\mathbf{r}} - e\phi(\mathbf{r})$$

The action is the relativistic arclength of the path of the particle in space time, minus the potential energy contribution, plus an extra contribution which quantum mechanically is an extra phase a charged particle gets when it is moving along a vector potential.

Derivation of Lorentz force from relativistic Lagrangian (SI units)

The equations of motion derived by extremizing the action (see matrix calculus for the notation):

$$\frac{\mathrm{d}\mathbf{P}}{\mathrm{d}t} = \frac{\partial L}{\partial \mathbf{r}} = e\frac{\partial \mathbf{A}}{\partial \mathbf{r}} \cdot \dot{\mathbf{r}} - e\frac{\partial \phi}{\partial \mathbf{r}}$$

$$\mathbf{P} - e\mathbf{A} = \frac{m\dot{\mathbf{r}}}{\sqrt{1 - \left(\frac{\dot{\mathbf{r}}}{c}\right)^2}}$$

are the same as Hamilton's equations of motion:

$$\frac{\mathrm{d}\mathbf{r}}{\mathrm{d}t} = \frac{\partial}{\partial \mathbf{p}}\left(\sqrt{(\mathbf{P} - e\mathbf{A})^2 + (mc^2)^2} + e\phi\right)$$

$$\frac{\mathrm{d}\mathbf{p}}{\mathrm{d}t} = -\frac{\partial}{\partial \mathbf{r}}\left(\sqrt{(\mathbf{P} - e\mathbf{A})^2 + (mc^2)^2} + e\phi\right)$$

both are equivalent to the noncanonical form:

$$\frac{\mathrm{d}}{\mathrm{d}t}\left(\frac{m\dot{\mathbf{r}}}{\sqrt{1 - \left(\frac{\dot{\mathbf{r}}}{c}\right)^2}}\right) = e\left(\mathbf{E} + \dot{\mathbf{r}} \times \mathbf{B}\right).$$

This formula is the Lorentz force, representing the rate at which the EM field adds relativistic momentum to the particle.

Equation (CGS Units)

The above-mentioned formulae use SI units which are the most common among experimentalists, technicians, and engineers. In cgs-Gaussian units, which are somewhat more common among theoretical physicists, one has instead

$$\mathbf{F} = q_{\text{cgs}}\left(\mathbf{E}_{\text{cgs}} + \frac{\mathbf{v}}{c} \times \mathbf{B}_{\text{cgs}}\right).$$

where c is the speed of light. Although this equation looks slightly different, it is completely equivalent, since one has the following relations:

$$q_{\text{cgs}} = \frac{q_{\text{SI}}}{\sqrt{4\pi\epsilon_0}}, \quad \mathbf{E}_{\text{cgs}} = \sqrt{4\pi\epsilon_0}\,\mathbf{E}_{\text{SI}}, \quad \mathbf{B}_{\text{cgs}} = \sqrt{4\pi/\mu_0}\,\mathbf{B}_{\text{SI}}$$

where ε_0 is the vacuum permittivity and μ_0 the vacuum permeability. In practice, the subscripts "cgs" and "SI" are always omitted, and the unit system has to be assessed from context.

RELATIVISTIC FORM OF THE LORENTZ FORCE

Covariant form of the Lorentz Force

Field Tensor

Using the metric signature (1, −1, −1, −1), The Lorentz force for a charge q can be written in covariant form:

$$\frac{dp^\alpha}{d\tau} = qF^{\alpha\beta}U_\beta$$

where pα is the four-momentum, defined as

$$p^\alpha = (p_0, p_1, p_2, p_3) = (\gamma mc, p_x, p_y, p_z),$$

τ the proper time of the particle, Fαβ the contravariant electromagnetic tensor

$$F^{\alpha\beta} = \begin{pmatrix} 0 & -E_x/c & -E_y/c & -E_z/c \\ E_x/c & 0 & -B_z & B_y \\ E_y/c & B_z & 0 & -B_x \\ E_z/c & -B_y & B_x & 0 \end{pmatrix}$$

and U is the covariant 4-velocity of the particle, defined as:

$$U_\beta = (U_0, U_1, U_2, U_3) = \gamma\,(c, -v_x, -v_y, -v_z)\ ,$$

in which

$$\gamma(v) = \frac{1}{\sqrt{1-\frac{v^2}{c^2}}} = \frac{1}{\sqrt{1-\frac{v_x^2+v_y^2+v_z^2}{c^2}}}$$

is the Lorentz factor.

The fields are transformed to a frame moving with constant relative velocity by:

$$F'^{\mu\nu} = \Lambda^{\mu}{}_{\alpha}\Lambda^{\nu}{}_{\beta}F^{\alpha\beta}\ ,$$

where Λμα is the Lorentz transformation tensor.

Translation to Vector Notation

The α = 1 component (x-component) of the force is

The α = 1 component (*x*-component) of the force is

$$\frac{\mathrm{d}p^1}{\mathrm{d}\tau} = qU_\beta F^{1\beta} = q\left(U_0F^{10} + U_1F^{11} + U_2F^{12} + U_3F^{13}\right).$$

Substituting the components of the covariant electromagnetic tensor *F* yields

Using the components of covariant four-velocity yields

$$\begin{aligned}\frac{\mathrm{d}p^1}{\mathrm{d}\tau} &= q\gamma\left[c\left(\frac{E_x}{c}\right) + (-v_y)(-B_z) + (-v_z)(B_y)\right]\\ &= q\gamma\left(E_x + v_yB_z - v_zB_y\right)\\ &= q\gamma\left[E_x + (\mathbf{v}\times\mathbf{B})_x\right].\end{aligned}$$

The calculation for α = 2, 3 (force components in the y and z directions) yields similar results, so collecting the 3 equations into one:

$$\frac{\mathrm{d}\mathbf{p}}{\mathrm{d}\tau} = q\gamma\left(\mathbf{E} + \mathbf{v}\times\mathbf{B}\right),$$

and since differentials in coordinate time *dt* and proper time *dτ* are related by the Lorentz factor,

$$dt = \gamma(v)d\tau\,,$$

so we arrive at

$$\frac{\mathrm{d}\mathbf{p}}{\mathrm{d}t} = q\left(\mathbf{E} + \mathbf{v}\times\mathbf{B}\right).$$

This is precisely the Lorentz force law, however, it is important to note that **p** is the relativistic expression,

$$\mathbf{p} = \gamma(v)m_0\mathbf{v}\,.$$

Lorentz Force in Spacetime Algebra (STA)

The electric and magnetic fields are dependent on the velocity of an observer, so the relativistic form of the Lorentz force law can best be exhibited starting from a coordinate-independent expression for the electromagnetic and magnetic fields $\mathcal{F}$, and an arbitrary time-direction, γ_0. This can be settled through Space-Time Algebra (or the geometric algebra of space-time), a type of Clifford's Algebra defined on a pseudo-euclidian space,as

$$\mathbf{E} = (\mathcal{F} \cdot \gamma_0)\gamma_0$$

and

$$i\mathbf{B} = (\mathcal{F} \wedge \gamma_0)\gamma_0$$

F is a space-time bivector (an oriented plane segment, just like a vector is an oriented line segment), which has six degrees of freedom corresponding to boosts (rotations in space-time planes) and rotations (rotations in space-space planes). The dot product with the vector γ_0 pulls a vector (in the space algebra) from the translational part, while the wedge-product creates a trivector (in the space algebra) who is dual to a vector which is the usual magnetic field vector. The relativistic velocity is given by the (time-like) changes in a time-position vector $v = \dot{x}$, where

$$v^2 = 1,$$

(which shows our choice for the metric) and the velocity is

$$\mathbf{v} = cv \wedge \gamma_0 / (v \cdot \gamma_0).$$

The proper (invariant is an inadequate term because no transformation has been defined) form of the Lorentz force law is simply

$$F = q\mathcal{F} \cdot v$$

Note that the order is important because between a bivector and a vector the dot product is anti-symmetric. Upon a space time split like one can obtain the velocity, and fields as above yielding the usual expression.

Applications

The Lorentz force occurs in many devices, including:

- Cyclotrons and other circular path particle accelerators
- Mass spectrometers
- Velocity Filters
- Magnetrons
- Lorentz force velocimetry

In its manifestation as the Laplace force on an electric current in a conductor, this force occurs in many devices including:

- Electric motors
- Railguns
- Linear motors
- Loudspeakers
- Magnetoplasmadynamic thrusters
- Electrical generators
- Homopolar generators
- Linear alternators

ELECTROMAGNETIC INDUCTION

Electromagnetic or Magnetic induction is the production of an electromotive force (i.e., voltage) across an electrical conductor due to its dynamic interaction with a magnetic field.

Michael Faraday is generally credited with the discovery of induction in 1831, and James Clerk Maxwell mathematically described it as Faraday's law of induction. Lenz's law describes the direction of the induced field. Faraday's law was later generalized to the Maxwell-Faraday equation, which is one of the equations in James Clerk Maxwell's theory of electromagnetism.

Electromagnetic induction has found many applications in technology, including electrical components such as inductors and transformers, and devices such as electric motors and generators.

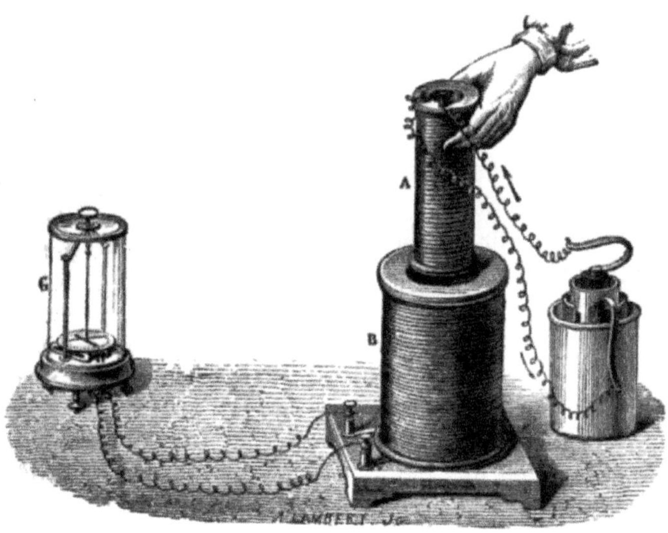

Faraday's experiment showing induction between coils of wire: The liquid battery (right) provides a current that flows through the small coil (A), creating a magnetic field. When the coils are stationary, no current is induced. But when the small coil is moved in or out of the large coil (B), the magnetic flux through the large coil changes, inducing a current which is detected by the galvanometer (G).

Electromagnetic induction was first discovered by Michael Faraday, who made his discovery public in 1831. It was discovered independently by Joseph Henry in 1832.

In Faraday's first experimental demonstration (August 29, 1831), he wrapped two wires around opposite sides of an iron ring or "torus" (an arrangement similar to a modern toroidal transformer). Based on his understanding of electromagnets, he expected that, when current started to flow in one wire, a sort of wave would travel through the ring and cause some electrical effect on the opposite side. He plugged one wire into a galvanometer, and watched it as he connected the other wire to a battery. He saw a transient current, which he called a "wave of electricity", when he connected the wire to the battery and another when he disconnected it. This induction was due to the change in magnetic flux that occurred when the battery was connected and disconnected. Within two months, Faraday found several other manifestations of electromagnetic induction. For example, he saw transient currents when he quickly slid a bar magnet in and out of a coil of wires, and he generated a steady (DC) current by rotating a copper disk near the bar magnet with a sliding electrical lead ("Faraday's disk").

Faraday explained electromagnetic induction using a concept he called lines of force. However, scientists at the time widely rejected his theoretical ideas, mainly because they were not formulated mathematically. An exception was James Clerk Maxwell, who used Faraday's ideas as the basis of his quantitative electromagnetic theory. In Maxwell's model, the time varying aspect of electromagnetic induction is expressed as a differential equation, which Oliver Heaviside referred to as Faraday's law even though it is slightly different from Faraday's original formulation and does not describe motional EMF. Heaviside's version (see Maxwell–Faraday equation below) is the form recognized today in the group of equations known as Maxwell's equations.

In 1834 Heinrich Lenz formulated the law named after him to describe the "flux through the circuit". Lenz's law gives the direction of the induced EMF and current resulting from electromagnetic induction.

A diagram of Faraday's iron ring apparatus. Change in the magnetic flux of the left coil induces a current in the right coil.

Faraday's disk (see homopolar generator)

Theory

Faraday's Law of Induction and Lenz's Law

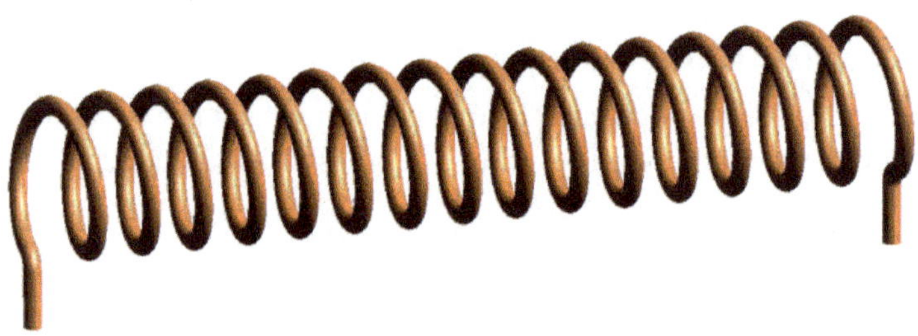

A solenoid

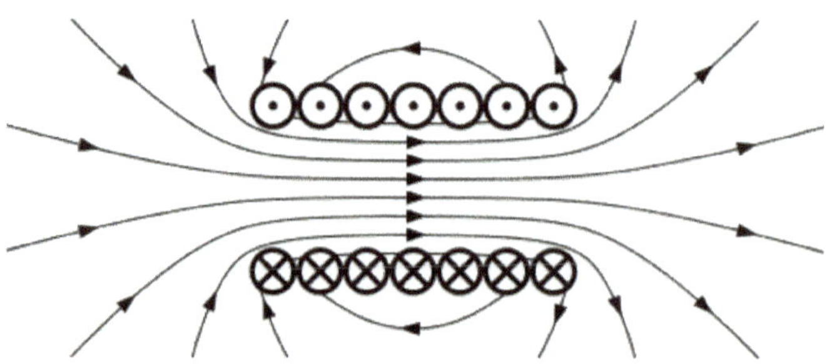

The longitudinal cross section of a solenoid with a constant electrical current running through it. The magnetic field lines are indicated, with their direction shown by arrows.

The magnetic flux corresponds to the 'density of field lines'. The magnetic flux is thus densest in the middle of the solenoid, and weakest outside of it.

Faraday's law of induction makes use of the magnetic flux Φ_B through a region of space enclosed by a wire loop. The magnetic flux is defined by a surface integral:

$$\Phi_B = \int_{\Sigma} \mathbf{B} \cdot d\mathbf{A} \,,$$

where dA is an element of the surface Σ enclosed by the wire loop, B is the magnetic field. The dot product B·dA corresponds to an infinitesimal amount of magnetic flux. In more visual terms, the magnetic flux through the wire loop is proportional to the number of magnetic flux lines that pass through the loop.

When the flux through the surface changes, Faraday's law of induction says that the wire loop acquires an electromotive force (EMF).[note 1] The most widespread version of this law states that the induced electromotive force in any closed circuit is equal to the rate of change of the magnetic flux enclosed by the circuit:

$$\mathcal{E} = -\frac{d\Phi_B}{dt} \,,$$

Where $\mathcal{E}$ is the EMF and Φ_B is the magnetic flux. The direction of the electromotive force is given by Lenz's law which states that an induced current will flow in the direction that will oppose the change which produced it. This is due to the negative sign in the previous equation. To increase the generated EMF, a common approach is to exploit flux linkage by creating a tightly wound coil of wire, composed of N identical turns, each with the same magnetic flux going through them. The resulting EMF is then N times that of one single wire.

$$\mathcal{E} = -N\frac{d\Phi_B}{dt}$$

Generating an EMF through a variation of the magnetic flux through the surface of a wire loop can be achieved in several ways:

- the magnetic field **B** changes (e.g. an alternating magnetic field, or moving a wire loop towards a bar magnet where the B field is stronger),
- the wire loop is deformed and the surface Σ changes,
- the orientation of the surface d**A** changes (e.g. spinning a wire loop into a fixed magnetic field),
- any combination of the above

Maxwell–Faraday Equation

In general, the relation between the EMF $\mathcal{E}$ in a wire loop encircling a surface Σ, and the electric field **E** in the wire is given by

$$\mathcal{E} = \oint_{\partial \Sigma} \mathbf{E} \cdot d\boldsymbol{\ell}$$

where $d\boldsymbol{\ell}$ is an element of contour of the surface Σ, combining this with the definition of flux

$$\Phi_B = \int_{\Sigma} \mathbf{B} \cdot d\mathbf{A},$$

we can write the integral form of the Maxwell–Faraday equation

$$\oint_{\partial \Sigma} \mathbf{E} \cdot d\boldsymbol{\ell} = -\frac{d}{dt} \int_{\Sigma} \mathbf{B} \cdot d\mathbf{A}$$

It is one of the four Maxwell's equations, and therefore plays a fundamental role in the theory of classical electromagnetism.

Faraday's Law and Relativity

Faraday's law describes two different phenomena: the motional EMF generated by a magnetic force on a moving wire (see Lorentz force), and the transformer EMF generated by an electric force due to a changing magnetic field (due to the differential form of the Maxwell–Faraday equation). James Clerk Maxwell drew attention to the separate physical phenomena in 1861. This is believed to be a unique example in physics of where such a fundamental law is invoked to explain two such different phenomena.

Einstein noticed that the two situations both corresponded to a relative movement between a conductor and a magnet, and the outcome was unaffected by which one was moving. This was one of the principal paths that led him to develop special relativity.

Applications

The principles of electromagnetic induction are applied in many devices and systems, including:

- Current clamp
- Electric generators
- Electromagnetic forming
- Graphics tablet
- Hall effect meters
- Induction cooking
- Induction motors
- Induction sealing
- Induction welding
- Inductive charging
- Inductors

- Magnetic flow meters
- Mechanically powered flashlight
- Pickups
- Rowland ring
- Transcranial magnetic stimulation
- Transformers
- Wireless energy transfer

Electrical Generator

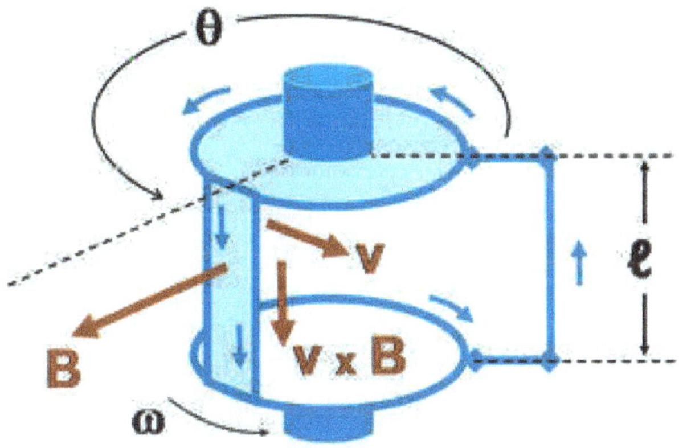

Rectangular wire loop rotating at angular velocity ω in radially outward pointing magnetic field **B** of fixed magnitude. The circuit is completed by brushes making sliding contact with top and bottom discs, which have conducting rims. This is a simplified version of the drum generator.

The EMF generated by Faraday's law of induction due to relative movement of a circuit and a magnetic field is the phenomenon underlying electrical generators. When a permanent magnet is moved relative to a conductor, or vice versa, an electromotive force is created. If the wire is connected through an electrical load, current will flow, and thus electrical energy is generated, converting the mechanical energy of motion to electrical energy. For example, the drum generator is based upon the figure to the right. A different implementation of this idea is the Faraday's disc, shown in simplified form on the right.

In the Faraday›s disc example, the disc is rotated in a uniform magnetic field perpendicular to the disc, causing a current to flow in the radial arm due to the Lorentz force. It is interesting to understand how it arises that mechanical work is necessary to drive this current. When the generated current flows through the conducting rim, a magnetic field is generated by this current through Ampère›s circuital law (labeled "induced B" in the figure). The rim thus becomes an electromagnet that resists rotation of the disc (an example of Lenz›s law). On the far side of the figure, the return current flows from the rotating arm through the far side of the rim to the bottom brush. The B-field induced by this return current opposes the applied B-field, tending to decrease the flux through that side of the circuit, opposing the increase in flux due to rotation. On the near side of the figure, the return current flows from the rotating arm through the near side of the rim to the bottom brush. The induced B-field increases the flux on this side of the circuit, opposing the decrease in flux due to rotation. Thus, both sides of the circuit generate an EMF opposing the rotation. The energy required to keep the disc moving, despite this reactive force, is exactly equal to the electrical energy generated (plus energy wasted due to friction, Joule heating, and other inefficiencies). This behavior is common to all generators converting mechanical energy to electrical energy.

Electrical Transformer

When the electric current in a loop of wire changes, the changing current creates a changing magnetic field. A second wire in reach of this magnetic field will experience this change in magnetic field as a change in its coupled magnetic flux, $d\,\Phi_B\,/\,d\,t$. Therefore, an electromotive force is set up in the second loop called the induced EMF or transformer EMF. If the two ends of this loop are connected through an electrical load, current will flow.

Current Clamp

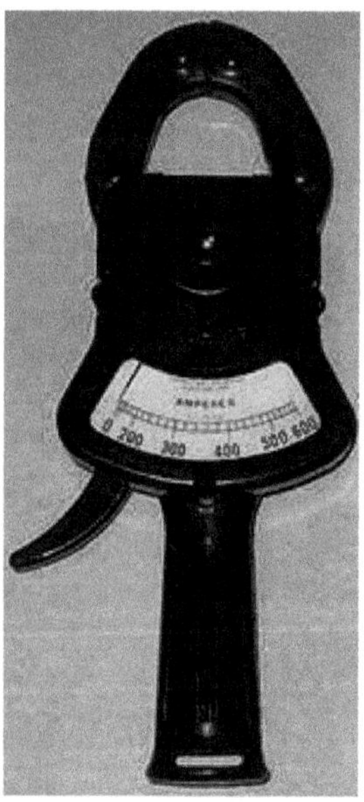

A current clamp

A current clamp is a type of transformer with a split core which can be spread apart and clipped onto a wire or coil to either measure the current in it or, in reverse, to induce a voltage. Unlike conventional instruments the clamp does not make electrical contact with the conductor or require it to be disconnected during attachment of the clamp.

Magnetic Flow Meter

Faraday's law is used for measuring the flow of electrically conductive liquids and slurries. Such instruments are called magnetic flow meters. The induced voltage $\mathcal{E}$ generated in the magnetic field B due to a conductive liquid moving at velocity v is thus given by:

$$\mathcal{E} = -B\ell v,$$

where ℓ is the distance between electrodes in the magnetic flow meter.

Eddy Currents

Conductors (of finite dimensions) moving through a uniform magnetic field, or stationary within a changing magnetic field, will have currents induced within them. These induced eddy currents can be undesirable, since they dissipate energy in the resistance of the conductor. There are a number of methods employed to control these undesirable inductive effects.

- Electromagnets in electric motors, generators, and transformers do not use solid metal, but instead use thin sheets of metal plate, called laminations. These thin plates reduce the parasitic eddy currents, as described below.
- Inductive coils in electronics typically use magnetic cores to minimize parasitic current flow. They are a mixture of metal powder plus a resin binder that can hold any shape. The binder prevents parasitic current flow through the powdered metal.

Electromagnet Laminations

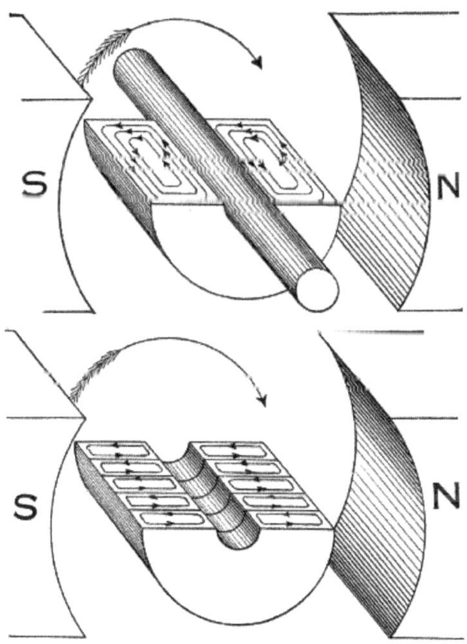

Eddy currents occur when a solid metallic mass is rotated in a magnetic field, because the outer portion of the metal cuts more lines of force than the inner portion, hence the induced electromotive force not being uniform, tends to set up currents between the points of greatest and least potential. Eddy currents consume a considerable amount of energy and often cause a harmful rise in temperature.

Only five laminations or plates are shown in this example, so as to show the subdivision of the eddy currents. In practical use, the number of laminations or punchings ranges from 40 to 66 per inch, and brings the eddy current loss down to about one percent. While the plates can be separated by insulation, the voltage is so low that the natural rust/oxide coating of the plates is enough to prevent current flow across the laminations.

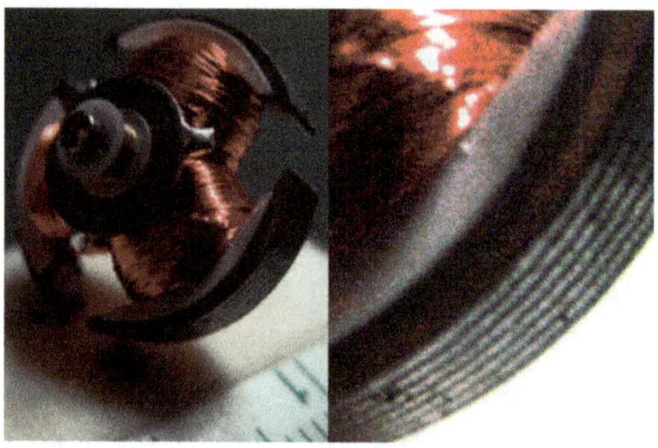

This is a rotor approximately 20mm in diameter from a DC motor used in a CD player. Note the laminations of the electromagnet pole pieces, used to limit parasitic inductive losses.

Parasitic Induction within Conductors

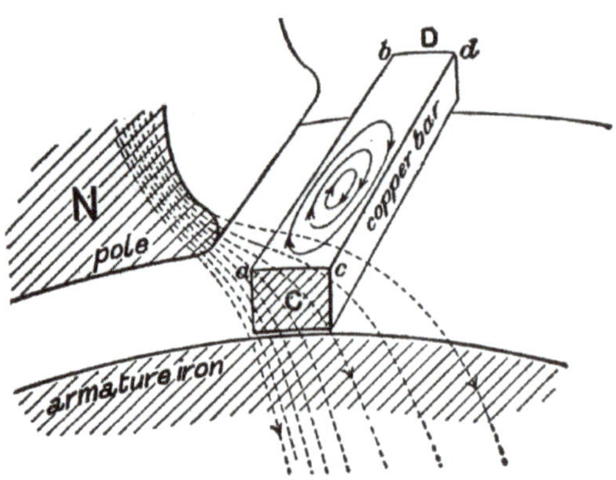

In this illustration, a solid copper bar conductor on a rotating armature is just passing under the tip of the pole piece N of the field magnet. Note the uneven distribution of the lines of force across the copper bar. The magnetic field is more concentrated and thus stronger on the left edge of the copper bar (a,b) while the field is weaker on the right edge (c,d). Since the two edges of the bar move with the same velocity, this difference in field strength across the bar creates whorls or current eddies within the copper bar.

High current power-frequency devices, such as electric motors, generators and transformers, use multiple small conductors in parallel to break up the eddy flows that can form within large solid conductors. The same principle is applied to transformers used at higher than power frequency, for example, those used in switch-mode power supplies and the intermediate frequency coupling transformers of radio receivers.

FARADAY'S LAW OF INDUCTION

Faraday's law of induction is a basic law of electromagnetism predicting how a magnetic field will interact with an electric circuit to produce an electromotive force (EMF)—a phenomenon called electromagnetic induction. It is the fundamental operating principle of transformers, inductors, and many types of electrical motors, generators and solenoids.

The Maxwell–Faraday equation is a generalization of Faraday's law, and is listed as one of Maxwell's equations.

Faraday›s Law

Qualitative Statement

The most widespread version of Faraday's law states:

The induced electromotive force in any closed circuit is equal to the negative of the time rate of change of the magnetic flux enclosed by the circuit.

This version of Faraday's law strictly holds only when the closed circuit is a loop of infinitely thin wire, and is invalid in other circumstances as discussed below. A different version, the Maxwell–Faraday equation (discussed below), is valid in all circumstances.

Quantitative

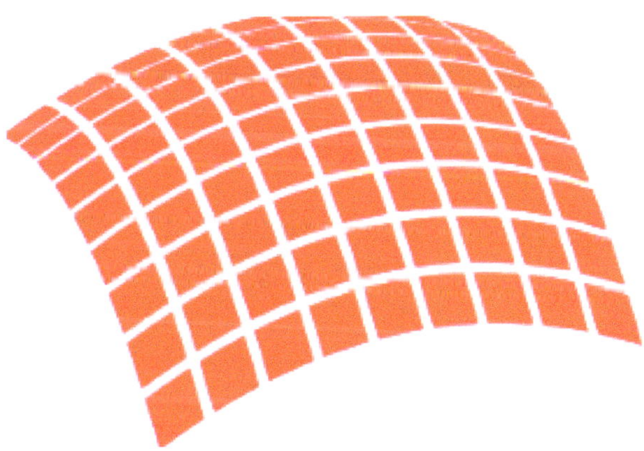

The definition of surface integral relies on splitting the surface Σ into small surface elements. Each element is associated with a vector **dA** of magnitude equal to the area of the element and with direction normal to the element and pointing "outward" (with respect to the orientation of the surface).

Faraday›s law of induction makes use of the magnetic flux Φ_B through a hypothetical surface Σ whose boundary is a wire loop. Since the wire loop may be moving, we write Σ(t) for the surface. The magnetic flux is defined by a surface integral:

$$\Phi_B = \iint_{\Sigma(t)} \mathbf{B}(\mathbf{r},t) \cdot d\mathbf{A},$$

where dA is an element of surface area of the moving surface Σ(t), B is the magnetic field (also called "magnetic flux density"), and B·dA is a vector dot product (the infinitesimal amount of magnetic flux through the infinitesimal area element dA). In more visual terms, the magnetic flux through the wire loop is proportional to the number of magnetic flux lines that pass through the loop.

When the flux changes—because B changes, or because the wire loop is moved or deformed, or both—Faraday's law of induction says that the wire loop acquires an EMF, $\mathcal{E}$, defined as the energy available from a unit charge that has travelled once around the wire loop. Equivalently, it is the voltage that would be measured by cutting the wire to create an open circuit, and attaching a voltmeter to the leads.

Faraday's law states that the EMF is also given by the rate of change of the magnetic flux:

$$\mathcal{E} = -\frac{d\Phi_B}{dt},$$

where $\mathcal{E}$ is the electromotive force (EMF) and Φ_B is the magnetic flux. The direction of the electromotive force is given by Lenz's law.

For a tightly wound coil of wire, composed of N identical turns, each with the same Φ_B, Faraday's law of induction states that

$$\mathcal{E} = -N\frac{d\Phi_B}{dt}$$

where N is the number of turns of wire and Φ_B is the magnetic flux through a single loop.

Maxwell–Faraday Equation

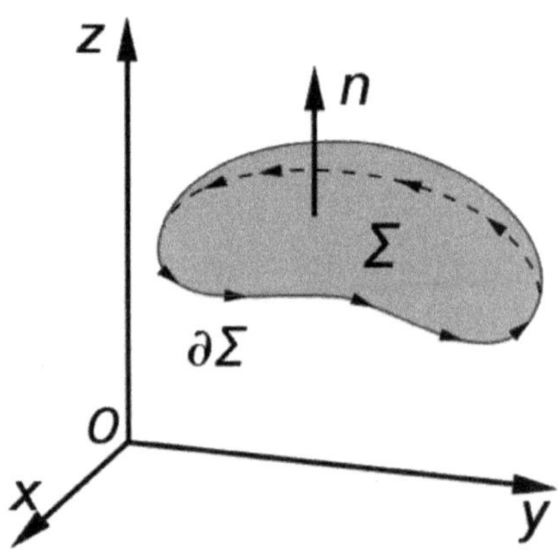

An illustration of the Kelvin–Stokes theorem with surface **Σ**, its boundary ∂**Σ**, and orientation **n** set by the right-hand rule.

The Maxwell–Faraday equation is a generalisation of Faraday›s law that states that a time-varying magnetic field will always accompany a spatially-varying, non-conservative electric field, and vice versa. The Maxwell–Faraday equation is

$$\nabla \times \mathbf{E} = -\frac{\partial \mathbf{B}}{\partial t}$$

(in SI units) where ∇ × is the curl operator and again **E**(**r**, t) is the electric field and **B**(**r**, t) is the magnetic field. These fields can generally be functions of position **r** and time t.

The Maxwell–Faraday equation is one of the four Maxwell's equations, and therefore plays a fundamental role in the theory of classical electromagnetism. It can also be written in an **integral form** by the Kelvin–Stokes theorem:

$$\oint_{\partial\Sigma} \mathbf{E} \cdot d\mathbf{l} = -\int_{\Sigma} \frac{\partial \mathbf{B}}{\partial t} \cdot d\mathbf{A}$$

Where

as indicated in the figure:

Σ is a surface bounded by the closed contour ∂**Σ**,

E is the electric field, **B** is the magnetic field.

d**l** is an infinitesimal vector element of the contour ∂**Σ**,

d**A** is an infinitesimal vector element of surface **Σ**. If its direction is orthogonal to that surface patch, the magnitude is the area of an infinitesimal patch of surface.

Both d**l** and d**A** have a sign ambiguity; to get the correct sign, the right-hand rule is used, as explained in the article Kelvin–Stokes theorem. For a planar surface **Σ**, a positive path element d**l** of curve ∂**Σ** is defined by the right-hand rule as one that points with the fingers of the right hand when the thumb points in the direction of the normal **n** to the surface **Σ**.

The integral around ∂**Σ** is called a path integral or line integral.

Notice that a nonzero path integral for **E** is different from the behavior of the electric field generated by charges. A charge-generated **E**-field can be expressed as the gradient of a scalar field that is a solution to Poisson's equation, and has a zero path integral. See gradient theorem.

The integral equation is true for any path ∂**Σ** through space, and any surface **Σ** for which that path is a boundary.

If the path **Σ** is not changing in time, the equation can be rewritten:

$$\oint_{\partial\Sigma} \mathbf{E} \cdot d\mathbf{l} = -\frac{d}{dt} \int_{\Sigma} \mathbf{B} \cdot d\mathbf{A}.$$

The surface integral at the right-hand side is the explicit expression for the magnetic flux Φ_B through Σ.

Proof of Faraday's Law

The four Maxwell's equations (including the Maxwell–Faraday equation), along with the Lorentz force law, are a sufficient foundation to derive everything in classical electromagnetism. Therefore, it is possible to "prove" Faraday's law starting with these equations. Click "show" in the box below for an outline of this proof. (In an alternative approach, not shown here but equally valid, Faraday›s law could be taken as the starting point and used to "prove" the Maxwell–Faraday equation and/or other laws.)

Consider the time-derivative of flux through a possibly moving loop, with area $\Sigma(t)$:

$$\frac{d\Phi_B}{dt} = \frac{d}{dt}\int_{\Sigma(t)} \mathbf{B}(t)\cdot d\mathbf{A}$$

The integral can change over time for two reasons: The integrand can change, or the integration region can change. These add linearly, therefore:

$$\left.\frac{d\Phi_B}{dt}\right|_{t=t_0} = \left(\int_{\Sigma(t_0)} \left.\frac{\partial \mathbf{B}}{\partial t}\right|_{t=t_0} \cdot d\mathbf{A}\right) + \left(\frac{d}{dt}\int_{\Sigma(t)} \mathbf{B}(t_0)\cdot d\mathbf{A}\right)$$

where t_0 is any given fixed time. We will show that the first term on the right-hand side corresponds to transformer EMF, the second to motional EMF (see above). The first term on the right-hand side can be rewritten using the integral form of the Maxwell–Faraday equation:

$$\int_{\Sigma(t_0)} \left.\frac{\partial \mathbf{B}}{\partial t}\right|_{t=t_0} \cdot d\mathbf{A} = -\oint_{\partial\Sigma(t_0)} \mathbf{E}(t_0)\cdot d\mathbf{l}$$

Next, we analyze the second term on the right-hand side:

$$\frac{d}{dt}\int_{\Sigma(t)} \mathbf{B}(t_0)\cdot d\mathbf{A}$$

This is the most difficult part of the proof; more details and alternate approaches can be found in references. As the loop moves and/or deforms, it sweeps out a surface (see figure on right). The magnetic flux through this swept-out surface corresponds to the magnetic flux that is either entering or exiting the loop, and therefore this is the magnetic flux that contributes to the time-derivative. (This step implicitly uses Gauss's law for magnetism: Since the flux lines have no beginning or end, they can only get into the loop by getting cut through by the wire.) As a small part of the loop d**l** moves with velocity **v** for a short time dt, it sweeps out a vector area vector d**A** = **v** dt × d**l**. Therefore, the change in magnetic flux through the loop here is

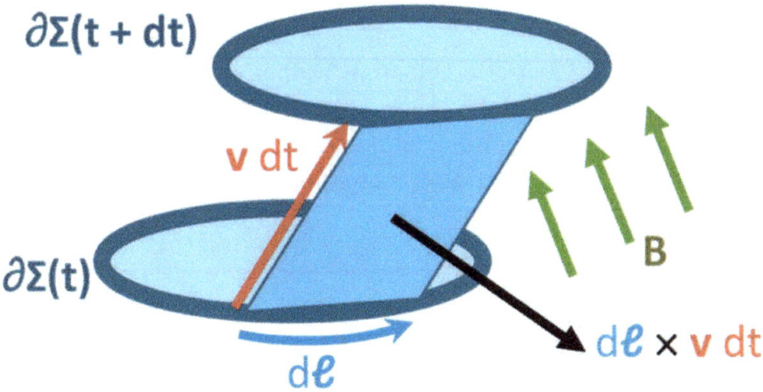

Area swept out by vector element dl of curve $\partial\Sigma$ in time dt when moving with velocity v.

$$\mathbf{B}\cdot(\mathbf{v}\,dt\times d\mathbf{l}) = -dt\,d\mathbf{l}\cdot(\mathbf{v}\times\mathbf{B})$$

Therefore:

$$\frac{d}{dt}\int_{\Sigma(t)}\mathbf{B}(t_0)\cdot d\mathbf{A} = -\oint_{\partial\Sigma(t_0)}(\mathbf{v}(t_0)\times\mathbf{B}(t_0))\cdot d\mathbf{l}$$

where $\mathbf{v}$ is the velocity of a point on the loop $\partial\Sigma$.

Putting these together,

$$\left.\frac{d\Phi_B}{dt}\right|_{t=t_0} = \left(-\oint_{\partial\Sigma(t_0)}\mathbf{E}(t_0)\cdot d\mathbf{l}\right) + \left(-\oint_{\partial\Sigma(t_0)}(\mathbf{v}(t_0)\times\mathbf{B}(t_0))\cdot d\mathbf{l}\right)$$

Meanwhile, the EMF is defined as the energy available per unit charge that travels once around the wire loop. Therefore, by the Lorentz force law,

$$\mathcal{E} = \oint(\mathbf{E}+\mathbf{v}\times\mathbf{B})\cdot d\mathbf{l}$$

Combining these,

$$\frac{d\Phi_B}{dt} = -\mathcal{E}$$

"Counterexamples" to Faraday's Law

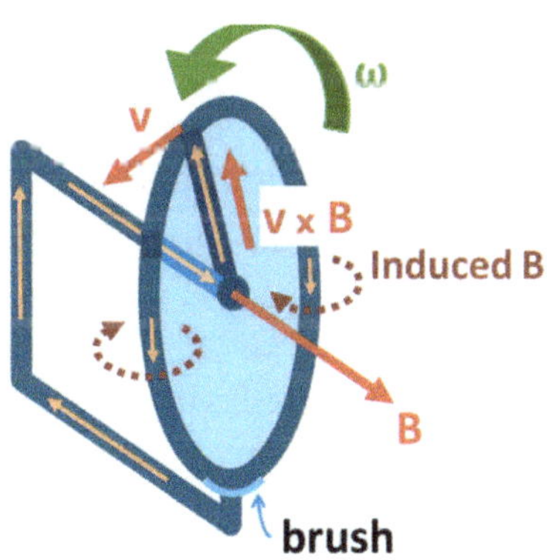

Faraday's disc electric generator. The disc rotates with angular rate ω, sweeping the conducting radius circularly in the static magnetic field **B**. The magnetic Lorentz force **v** × **B** drives the current along the conducting radius to the conducting rim, and from there the circuit completes through the lower brush and the axle supporting the disc. This device generates an EMF and a current, although the shape of the "circuit" is constant and thus the flux through the circuit does not change with time.

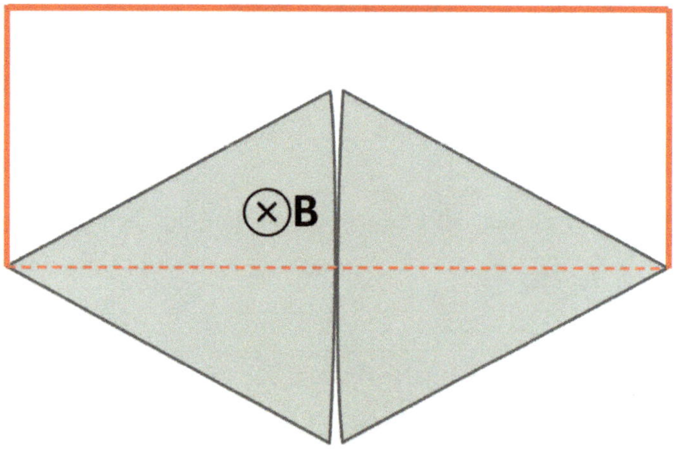

A counterexample to Faraday's law when over-broadly interpreted. A wire (solid red lines) connects to two touching metal plates (silver) to form a circuit. The whole system sits in a uniform magnetic field, normal to the page. If the word "circuit" is interpreted as "primary path of current flow" (marked in red), then the magnetic flux through the "circuit" changes dramatically as the plates are rotated, yet the EMF is almost zero, which contradicts Faraday's law. After Feynman Lectures on Physics Vol. II page 17-3.

Although Faraday›s law is always true for loops of thin wire, it can give the wrong result if naively extrapolated to other contexts. One example is the homopolar generator (above left): A spinning circular metal disc in a homogeneous magnetic field generates a DC (constant in time) EMF. In Faraday›s law, EMF is the time-derivative of flux, so a DC EMF is only possible if the magnetic flux is getting uniformly larger and larger perpetually. But in the generator, the magnetic field is constant and the disc stays in the same position, so no magnetic fluxes are growing larger and larger. So this example cannot be analyzed directly with Faraday›s law.

Another example, due to Feynman, has a dramatic change in flux through a circuit, even though the EMF is arbitrarily small. See figure and caption above right.

In both these examples, the changes in the current path are different from the motion of the material making up the circuit. The electrons in a material tend to follow the motion of the atoms that make up the material, due to scattering in the bulk and work function confinement at the edges. Therefore, motional EMF is generated when a material›s atoms are moving through a magnetic field, dragging the electrons with them, thus subjecting the electrons to the Lorentz force. In the homopolar generator, the material›s atoms are moving, even though the overall geometry of the circuit is staying the same. In the second example, the material›s atoms are almost stationary, even though the overall geometry of the circuit is changing dramatically. On the other hand, Faraday›s law always holds for thin wires, because there the geometry of the circuit always changes in a direct relationship to the motion of the material›s atoms.

Although Faraday›s law does not apply to all situations, the Maxwell–Faraday equation and Lorentz force law are always correct and can always be used directly.

Both of the above examples can be correctly worked by choosing the appropriate path of integration for Faraday›s law. Outside of context of thin wires, the path must never be chosen to go through the conductor in the shortest direct path.

Faraday's Law and Relativity

Two Phenomena

Some physicists have remarked that Faraday's law is a single equation describing two different phenomena: the motional EMF generated by a magnetic force on a moving wire (see Lorentz force), and the transformer EMF generated by an electric force due to a changing magnetic field (due to the Maxwell–Faraday equation).

James Clerk Maxwell drew attention to this fact in his 1861 paper On Physical Lines of Force. In the latter half of Part II of that paper, Maxwell gives a separate physical explanation for each of the two phenomena.

A reference to these two aspects of electromagnetic induction is made in some modern textbooks. As Richard Feynman states:

So the "flux rule" that the emf in a circuit is equal to the rate of change of the magnetic flux through the circuit applies whether the flux changes because the field changes or because the circuit moves (or both) ...

Yet in our explanation of the rule we have used two completely distinct laws for the two cases – $\mathbf{v} \times \mathbf{B}$ for "circuit moves" and $\nabla \times E = -\partial_t \mathbf{B}$ for "field changes".

We know of no other place in physics where such a simple and accurate general principle requires for its real understanding an analysis in terms of two different phenomena.

—Richard P. Feynman, The Feynman Lectures on Physics

Einstein's View

Reflection on this apparent dichotomy was one of the principal paths that led Einstein to develop special relativity:

It is known that Maxwell's electrodynamics—as usually understood at the present time—when applied to moving bodies, leads to asymmetries which do not appear to be inherent in the phenomena. Take, for example, the reciprocal electrodynamic action of a magnet and a conductor.

The observable phenomenon here depends only on the relative motion of the conductor and the magnet, whereas the customary view draws a sharp distinction between the two cases in which either the one or the other of these bodies is in motion. For if the magnet is in motion and the conductor at rest, there arises in the neighbourhood of the magnet an electric field with a certain definite energy, producing a current at the places where parts of the conductor are situated.

But if the magnet is stationary and the conductor in motion, no electric field arises in the neighbourhood of the magnet. In the conductor, however, we find an electromotive force, to which in itself there is no corresponding energy, but which gives rise—assuming equality of relative motion in the two cases discussed—to electric currents of the same path and intensity as those produced by the electric forces in the former case.

Examples of this sort, together with unsuccessful attempts to discover any motion of the earth relative to the "light medium," suggest that the phenomena of electrodynamics as well as of mechanics possess no properties corresponding to the idea of absolute rest.

REFERENCES

1. Ravaioli, Fawwaz T. Ulaby, Eric Michielssen, Umberto (2010). Fundamentals of applied electromagnetics (6th ed.). Boston: Prentice Hall. p. 13. ISBN 978-0-13-213931-1.
2. Martins, Roberto de Andrade. "Romagnosi and Volta's Pile: Early Difficulties in the Interpretation of Voltaic Electricity". In Fabio Bevilacqua and Lucio Fregonese (eds). Nuova Voltiana: Studies on Volta and his Times (PDF). vol. 3. Università degli Studi di Pavia. pp. 81–102. Retrieved 2010-12-02.
3. Stern, Dr. David P.; Peredo, Mauricio (2001-11-25). "Magnetic Fields -- History". NASA Goddard Space Flight Center. Retrieved 2009-11-27.
4. Feynman, Richard (1985). QED: The Strange Theory of Light and Matter. Princeton University Press. ISBN 978-0-691-12575-6.
5. Lamb, Willis; Retherford, Robert (1947). "Fine Structure of the Hydrogen Atom by a Microwave Method,". Physical Review. 72 (3): 241–243. Bibcode:1947PhRv...72..241L. doi:10.1103/PhysRev.72.241.
6. Foley, H.; Kusch, P. (1948). "On the Intrinsic Moment of the Electron". Physical Review. 73 (3): 412. Bibcode:1948PhRv...73..412F. doi:10.1103/PhysRev.73.412.
7. Barrett, Terence W. (2008). Topological Foundations of Electromagnetism. Singapore: World Scientific. p. 2. ISBN 978-981-277-997-7.
8. Ehrenberg, W; Siday, RE (1949). "The Refractive Index in Electron Optics and the Principles of Dynamics". Proceedings of the Physical Society. Series B. 62: 8–21. Bibcode:1949PPSB...62....8E. doi:10.1088/0370-1301/62/1/303.
9. Aharonov, Y; Bohm, D (1959). "Significance of electromagnetic potentials in quantum theory". Physical Review. 115: 485–491. Bibcode:1959PhRv..115..485A. doi:10.1103/PhysRev.115.485.
10. Al'tshuler,, B. L.; Aronov, A. G.; Spivak, B. Z. (1981). "The Aaronov-Bohm effect in disordered conductors" (PDF). Pis'ma Zh. Eksp. Teor. Fiz. 33: 101.
11. Berry, M. V. (1984). "Quantal Phase Factors Accompanying Adiabatic Changes" (PDF). Proc. Roy. Soc. A392 (1802): 45. doi:10.1098/rspa.1984.0023.

12. Aharonov, Y.; Anandan, J. (1987). "Phase change during a cyclic quantum evolution". Phys. Rev. Lett. 58 (16): 1593. doi:10.1103/PhysRevLett.58.1593.
13. Pancharatnam, S. (1956). "Generalized theory of interference, and its applications". Proceedings of the Indian Academy of Sciences. 44 (5): 247–262. doi:10.1007/BF03046050.
14. Chiao, Raymond Y.; Wu, Yong-Shi (1986). "Manifestations of Berry's Topological Phase for the Photon". Phys. Rev. Lett. 57 (8): 933. doi:10.1103/PhysRevLett.57.933.
15. B. D. Josephson (1962). "Possible new effects in superconductive tunnelling". Phys. Lett. 1 (7): 251–253. doi:10.1016/0031-9163(62)91369-0.
16. B. D. Josephson (1974). "The discovery of tunnelling supercurrents". Rev. Mod. Phys. 46 (2): 251–254. Bibcode:1974RvMP...46..251J. doi:10.1103/RevModPhys.46.251.
17. K. v. Klitzing; G. Dorda; M. Pepper (1980). "New method for high-accuracy determination of the fine-structure constant based on quantized Hall resistance". Phys. Rev. Lett. 45 (6): 494–497. Bibcode:1980PhRvL..45..494K. doi:10.1103/PhysRevLett.45.494.
18. de Haas, W. J.; van Alphen, P. M. (1930). "The dependance of the susceptibility of diamagnetic metals upon the field". Proc. Netherlands R. Acad. Sci. 33: 1106.
19. Penrose, Roger (2004). The Road to Reality (8th ed.). New York: Alfred A. Knopf. pp. 453–454. ISBN 0-679-45443-8.
20. Feng, J. H.; Kneubühl, F. K. (1995). Barrett, Terence William; Grimes, Dale M., eds. Solitons and Chaos in Periodic Nonlinear Optical Media and Lasers: Advanced Electromagnetism: Foundations, Theory and Applications. Singapore: World Scientific. p. 438. ISBN 981-02-2095-2.
21. Hunt, Bruce J. (2005). The Maxwellians. Cornell: Cornell University Press. p. 17. ISBN 978-0-8014-8234-2.
22. Josephs, H. J. (1959). "The Heaviside papers found at Paignton in 1957". The Institution of Electrical Engineers Monograph. 319: 70–76.
23. Hunt, Bruce J. (2005). The Maxwellians. Cornell: Cornell University Press. pp. 165–166. ISBN 978-0-8014-8234-2.
24. Barrett, T. W. (1991). "Tesla's Nonlinear Oscillator-Shuttle-Circuit (OSC) Theory" (PDF). Annales de la Fondation Louis de Broglie. 16 (1): 23–41. ISSN 0182-4295.
25. Penrose, Roger (2004). The Road to Reality (8th ed.). New York: Alfred A. Knopf. p. 201. ISBN 0-679-45443-8.
26. Rocher, E. Y. (1972). "Noumenon: Elementary entity of a new mechanics". J. Math. Phys. 13: 1919. doi:10.1063/1.1665933.
27. Imaeda, K. (1976). "A new formulation of classical electrodynamics". Il Nuovo Cimento B. 32 (1): 138–162. doi:10.1007/BF02726749.
28. Kauffmann, T.; Sun, Wen IyJ (1993). "Quaternion mechanics and electromagnetism.". Annales de la Fondation Louis de Broglie. 18 (2): 213–219.

29. Bostick, W. H.; Prior, W.; Grunberger, L.; Emmert, G. (1966). "Pair Production of Plasma Vortices". Physics of Fluids. 9 (10): 2078. doi:10.1063/1.1761572.
30. Ferraro, V .; Plumpton, C. (1961). An Introduction to Magneto-Fluid Mechanics. Oxford: Oxford University Press.
31. White, Carol (1977). Energy Potential: Toward a New Electromagnetic Field Theory (PDF). Campaigner Publications. p. 3. ISBN 0-918388-04-X.
32. Noether E (1918). "Invariante Variationsprobleme". Nachr. D. König. Gesellsch. D. Wiss. Zu Göttingen, Math-phys. Klasse. 1918: 235–257.
33. Scott, Alwyn (2006). Encyclopedia of Nonlinear Science. Routledge. p. 163. ISBN 978-1-135-45558-3.
34. Pismen, L. M. (2006). Patterns and Interfaces in Dissipative Dynamics. Springer. p. 3. ISBN 978-3-540-30431-9.
35. Kais, Sabre (2011). Popelier, Paul, ed. Finite Size Scaling for Criticality of the Schrodinger Equation: Solving the Schrodinger Equation: Has Everything Been Tried?. Singapore: Imperial College Press. pp. 91–92. ISBN 978-1-84816-724-7.
36. "Classical Physics Makes a Comeback". London: The Times. Jan 14, 1982.
37. Bostick, W. (1985). "Controversy over whether classical systems like plasmas can behave like superconductors (which have heretofore been supposed to be strictly quantum-mechanically dominated)". International Journal of Fusion Energy. 3 (2): 47–51. ISSN 0146-4981.
38. Bostick, W. (1985). "The morphology of the electron". International Journal of Fusion Energy. 3 (1): 9–52.
39. Bostick, W. (1985). "The morphology of the electron". International Journal of Fusion Energy. 3 (1): 68.
40. Edwards, W. Farrell (1981). "Classical Derivation of the London Equations". Phys. Rev. Lett. 47 (26): 1863. doi:10.1103/PhysRevLett.47.1863.
41. Fröhlich, H (1966). "Macroscopic wave functions in superconductors". Proceedings of the Physical Society. 87: 330. doi:10.1088/0370-1328/87/1/137.
42. Reed, Donald (1995). Barrett, Terence William; Grimes, Dale M., eds. Foundational Electrodynamics and Beltrami Vector Fields: Advanced Electromagnetism: Foundations, Theory and Applications. Singapore: World Scientific. p. 226. ISBN 981-02-2095-2.
43. Chen, Xie; Gu, Zheng-Cheng; Wen, Xiao-Gang (2010). "Local unitary transformation, long-range quantum entanglement, wave function renormalization, and topological order" (PDF). Phys. Rev. B. 82 (15): 155138. arXiv:1004.3835 . doi:10.1103/PhysRevB.82.155138.
44. International Union of Pure and Applied Chemistry (1993). Quantities, Units and Symbols in Physical Chemistry, 2nd edition, Oxford: Blackwell Science. ISBN 0-632-03583-8. pp. 14–15. Electronic version.
45. Oliver Heaviside By Paul J. Nahin, p120
46. Huray, Paul G. (2009). Maxwell's Equations. Wiley-IEEE. p. 22. ISBN 0-470-54276-4.

47. See Jackson page 2. The book lists the four modern Maxwell's equations, and then states, "Also essential for consideration of charged particle motion is the Lorentz force equation, F = q (E+ v × B), which gives the force acting on a point charge q in the presence of electromagnetic fields."

48. See Griffiths page 204.

49. For example, see the website of the "Lorentz Institute": \, or Griffiths.

50. Griffiths, David J. (1999). Introduction to electrodynamics. reprint. with corr. (3rd ed.). Upper Saddle River, New Jersey [u.a.]: Prentice Hall. ISBN 978-0-13-805326-0.

51. Delon, Michel (2001). Encyclopedia of the Enlightenment. Chicago, IL: Fitzroy Dearborn Publishers. p. 538. ISBN 157958246X.

52. Goodwin, Elliot H. (1965). The New Cambridge Modern History Volume 8: The American and French Revolutions, 1763–93. Cambridge: Cambridge University Press. p. 130. ISBN 9780521045469.

53. Meyer, Herbert W. (1972). A History of Electricity and Magnetism. Norwalk, Connecticut: Burndy Library. pp. 30 31. ISBN 0-262-13070-X.

54. Verschuur, Gerrit L. (1993). Hidden Attraction : The History And Mystery Of Magnetism. New York: Oxford University Press. pp. 78–79. ISBN 0-19-506488-7.

55. Darrigol, Olivier (2000). Electrodynamics from Ampère to Einstein. Oxford, [England]: Oxford University Press. pp. 9, 25. ISBN 0-19-850593-0.

56. Verschuur, Gerrit L. (1993). Hidden Attraction : The History And Mystery Of Magnetism. New York: Oxford University Press. p. 76. ISBN 0-19-506488-7.

57. Darrigol, Olivier (2000). Electrodynamics from Ampère to Einstein. Oxford, [England]: Oxford University Press. pp. 126–131, 139–144. ISBN 0-19-850593-0.

58. Darrigol, Olivier (2000). Electrodynamics from Ampère to Einstein. Oxford, [England]: Oxford University Press. pp. 200, 429–430. ISBN 0-19-850593-0.

59. Heaviside, Oliver. "On the Electromagnetic Effects due to the Motion of Electrification through a Dielectric". Philosophical Magazine, April 1889, p. 324.

60. Darrigol, Olivier (2000). Electrodynamics from Ampère to Einstein. Oxford, [England]. Oxford University Press. p. 327. ISBN 0-19-850593-0.

61. Whittaker, E. T. (1910). A History of the Theories of Aether and Electricity: From the Age of Descartes to the Close of the Nineteenth Century. Longmans, Green and Co. pp. 420–423. ISBN 1-143-01208-9.

62. See Griffiths page 326, which states that Maxwell's equations, "together with the [Lorentz] force law...summarize the entire theoretical content of classical electrodynamics".

63. for example, Jackson pp 777–8.

64. J.A. Wheeler; C. Misner; K.S. Thorne (1973). Gravitation. W.H. Freeman & Co. pp. 72–73. ISBN 0-7167-0344-0.. These authors use the Lorentz force in tensor form as definer of the electromagnetic tensorF, in turn the fields E and B.

65. I.S. Grant; W.R. Phillips; Manchester Physics (2008). Electromagnetism (2nd ed.). John Wiley & Sons. p. 122. ISBN 978-0-471-92712-9.
66. I.S. Grant; W.R. Phillips; Manchester Physics (2008). Electromagnetism (2nd Edition). John Wiley & Sons. p. 123. ISBN 978-0-471-92712-9.
67. Griffiths pages 301–3.
68. Tai L. Chow (2006). Electromagnetic theory. Sudbury MA: Jones and Bartlett. p. 395. ISBN 0-7637-3827-1.
69. Landau, L. D., Lifshits, E. M., & Pitaevskiĭ, L. P. (1984). Electrodynamics of continuous media; Volume 8 Course of Theoretical Physics (Second ed.). Oxford: Butterworth-Heinemann. p. §63 (§49 pp. 205–207 in 1960 edition). ISBN 0-7506-2634-8.
70. Roger F Harrington (2003). Introduction to electromagnetic engineering. Mineola, New York: Dover Publications. p. 56. ISBN 0-486-43241-6.
71. M N O Sadiku (2007). Elements of electromagnetics (Fourth ed.). NY/Oxford: Oxford University Press. p. 391. ISBN 0-19-530048-3.
72. Classical Mechanics (2nd Edition), T.W.B. Kibble, European Physics Series, Mc Graw Hill (UK), 1973, ISBN 0-07-084018-0.
73. Jackson, J.D. Chapter 11
74. Hestenes, David. "SpaceTime Calculus".
75. Poyser, A. W. (1892). Magnetism and Electricity: A Manual for Students in Advanced Classes. London and New York: Longmans, Green, & Co. p. 285.
76. Giancoli, Douglas C. (1998). Physics: Principles with Applications (Fifth ed.). pp. 623–624.
77. Ulaby, Fawwaz (2007). Fundamentals of applied electromagnetics (5th ed.). Pearson:Prentice Hall. p. 255. ISBN 0-13-241326-4.
78. "Joseph Henry". Distinguished Members Gallery, National Academy of Sciences. Retrieved 2006-11-30.
79. "A Brief History of Electromagnetism"(PDF).
80. "Electromagnetism". Smithsonian Institution Archives.
81. Michael Faraday, by L. Pearce Williams, p. 182-3
82. Michael Faraday, by L. Pearce Williams, p. 191–5
83. Michael Faraday, by L. Pearce Williams, p. 510
84. Maxwell, James Clerk (1904), A Treatise on Electricity and Magnetism, Vol. II, Third Edition. Oxford University Press, pp. 178–9 and 189.
85. "Archives Biographies: Michael Faraday", The Institution of Engineering and Technology.
86. Good, R. H. (1999). Classical Electromagnetism. Saunders College Publishing. p. 107. ISBN 0-03-022353-9.
87. Feynman, R. P.; Leighton, R. B.; Sands, M. L. (2006). The Feynman Lectures on Physics, Volume 2. Pearson/Addison-Wesley. pp. 17–2. ISBN 0-8053-9049-9.

88. Griffiths, D. J. (1999). Introduction to Electrodynamics (3rd ed.). Prentice Hall. pp. 301–303. ISBN 0-13-805326-X.
89. Tipler, P. A.; Mosca, G. (2003). Physics for Scientists and Engineers (5th ed.). W.H. Freeman. p. 795. ISBN 978-0716708100.
90. Jordan, E.; Balmain, K. G. (1968). Electromagnetic Waves and Radiating Systems (2nd ed.). Prentice-Hall. p. 100.
91. Hayt, W. (1989). Engineering Electromagnetics (5th ed.). McGraw-Hill. p. 312. ISBN 0-07-027406-1.
92. Schmitt, R. (2002). Electromagnetics Explained. p. 75.
93. Whelan, P. M.; Hodgeson, M. J. (1978). Essential Principles of Physics (2nd ed.). John Murray. ISBN 0-7195-3382-1.
94. Nave, C. R. "Faraday's Law". HyperPhysics. Georgia State University. Retrieved 2011-08-29.
95. Maxwell, J. C. (1861). "On physical lines of force". Philosophical Magazine. 90: 11–23. doi:10.1080/14786431003659180.
96. Griffiths, D. J. (1999). Introduction to Electrodynamics (3rd ed.). Prentice Hall. pp. 301–303. ISBN 0-13-805326-X. Note that the law relating flux to EMF, which this article calls "Faraday's law", is referred to by Griffiths as the "universal flux rule". He uses the term "Faraday's law" to refer to what this article calls the "Maxwell–Faraday equation".
97. "The flux rule" is the terminology that Feynman uses to refer to the law relating magnetic flux to EMF. Feynman, R. P.; Leighton, R. B.; Sands, M. L. (2006). The Feynman Lectures on Physics, Volume II. Pearson/Addison-Wesley. p. 17-2. ISBN 0-8053-9049-9.
98. Einstein, A. (1905). "Zur Elektrodynamik bewegter Körper". Annalen der Physik. 17 (10): 891–921. Bibcode:1905AnP...322..891E. doi:10.1002/andp.19053221004.
99. Translated in Einstein, A. (1923). "On the Electrodynamics of Moving Bodies"(PDF). The Principle of Relativity. Jeffery, G.B.; Perret, W. (transl.). London: Methuen and Company.
100. Images and reference text are from the public domain book: Hawkins Electrical Guide, Volume 1, Chapter 19: Theory of the Armature, pp. 270–273, Copyright 1917 by Theo. Audel & Co., Printed in the United States
101. Sadiku, M. N. O. (2007). Elements of Electromagnetics (4th ed.). New York & Oxford: Oxford University Press. p. 386. ISBN 0-19-530048-3.
102. "Applications of electromagnetic induction". Boston University. 1999-07-22.
103. Giancoli, Douglas C. (1998). Physics: Principles with Applications (Fifth ed.). pp. 623–624.
104. "A Brief History of Electromagnetism"(PDF).
105. Ulaby, Fawwaz (2007). Fundamentals of applied electromagnetics (5th ed.). Pearson:Prentice Hall. p. 255. ISBN 0-13-241326-4.

106. "Joseph Henry". Distinguished Members Gallery, National Academy of Sciences. Retrieved 2006-11-30.

107. Faraday, Michael; Day, P. (1999-02-01). The philosopher's tree: a selection of Michael Faraday's writings. CRC Press. p. 71. ISBN 978-0-7503-0570-9. Retrieved 28 August 2011.

108. Michael Faraday, by L. Pearce Williams, pp. 182–3

109. Michael Faraday, by L. Pearce Williams, p. 191–5

110. Michael Faraday, by L. Pearce Williams, p. 510

111. Maxwell, James Clerk (1904), A Treatise on Electricity and Magnetism, Vol. II, Third Edition. Oxford University Press, pp. 178–9 and 189.

112. "Archives Biographies: Michael Faraday", The Institution of Engineering and Technology.

113. Poyser, Arthur William (1892), Magnetism and electricity: A manual for students in advanced classes. London and New York; Longmans, Green, & Co., p. 285, fig. 248. Retrieved 2009-08-06.

114. "Faraday's Law, which states that the electromotive force around a closed path is equal to the negative of the time rate of change of magnetic flux enclosed by the path"Jordan, Edward; Balmain, Keith G. (1968). Electromagnetic Waves and Radiating Systems (2nd ed.). Prentice-Hall. p. 100.

115. "The magnetic flux is that flux which passes through any and every surface whose perimeter is the closed path"Hayt, William (1989). Engineering Electromagnetics (5th ed.). McGraw-Hill. p. 312. ISBN 0-07-027406-1.

116. The flux rule" is the terminology that Feynman uses to refer to the law relating magnetic flux to EMF. Feynman, R. P. (2006). Leighton, R. B.; Sands, M. L., eds. The Feynman Lectures on Physics. San Francisco: Pearson/Addison-Wesley. Vol. II, p. 17-2. ISBN 0-8053-9049-9.

117. Griffiths, David J. (1999). Introduction to Electrodynamics (Third ed.). Upper Saddle River NJ: Prentice Hall. pp. 301–303. ISBN 0-13-805326-X.

118. Tipler; Mosca. Physics for Scientists and Engineers. p. 795.

119. Note that different textbooks may give different definitions. The set of equations used throughout the text was chosen to be compatible with the special relativity theory.

120. Whelan, P. M.; Hodgeson, M. J. (1978). Essential Principles of Physics (2nd ed.). John Murray. ISBN 0-7195-3382-1.

121. Nave, Carl R. "Faraday's Law". HyperPhysics. Georgia State University. Retrieved 29 August 2011.

122. Harrington, Roger F. (2003). Introduction to electromagnetic engineering. Mineola, NY: Dover Publications. p. 56. ISBN 0-486-43241-6.

123. Davison, M. E. (1973). "A Simple Proof that the Lorentz Force, Law Implied Faraday's Law of Induction, when B is Time Independent". American Journal of Physics. 41 (5): 713–711. Bibcode:1973AmJPh..41..713D. doi:10.1119/1.1987339.

124. Krey; Owen. Basic Theoretical Physics: A Concise Overview. p. 155.

125. Simonyi, K. (1973). Theoretische Elektrotechnik (5th ed.). Berlin: VEB Deutscher Verlag der Wissenschaften. eq. 20, p. 47.

126. Hughes, W. F.; Young, F. J. (1965). The Electromagnetodynamics of Fluid. John Wiley.

127. Clerk Maxwell, James (1861). "On physical lines of force". Philosophical Magazine. Taylor & Francis. 90: 11–23. doi:10.1080/1478643100365918.

128. Griffiths, David J. (1999). Introduction to Electrodynamics (3rd ed.). Upper Saddle River, NJ: Prentice Hall. pp. 301–3. ISBN 0-13-805326-X. Note that the law relating flux to EMF, which this article calls "Faraday's law", is referred to in Griffiths' terminology as the "universal flux rule". Griffiths uses the term "Faraday's law" to refer to what this article calls the "Maxwell–Faraday equation". So in fact, in the textbook, Griffiths' statement is about the "universal flux rule".

129. Einstein, Albert. "On the Electrodynamics of Moving Bodies" (PDF).

CHAPTER 4

Power Engineering Basics

Chapter Contents

A "Power Engineer" is a technically skilled and knowledgeable professional who is certified and responsible to safely and efficiently operate equipment and processes that are regulated by boiler and pressure vessel legislation. Power Engineers are also known as: Steam Engineer. Stationary Engineer.

INTRODUCTION

Power engineering, also called power systems engineering, is a subfield of energy engineering and electrical engineering that deals with the generation, transmission, distribution and utilization of electric power and the electrical devices connected to such systems including generators, motors and transformers. Although much of the field is concerned with the problems of three-phase AC power – the standard for large-scale power transmission and distribution across the modern world – a significant fraction of the field is concerned with the conversion between AC and DC power and the development of specialized power systems such as those used in aircraft or for electric railway networks. Power engineering draws the majority of its theoretical base from electrical engineering and while some power engineers could be considered energy engineers, energy engineers often do not have the theoretical electrical engineering background to understand power engineering.

A steam turbine used to provide electric power.

A sketch of the Pearl Street Station, the first steam-powered electric power station in New York City

Electricity became a subject of scientific interest in the late 17th century with the work of William Gilbert. Over the next two centuries a number of important discoveries were made including the incandescent light bulb and the voltaic pile. Probably the greatest discovery with respect to power engineering came from Michael Faraday who in 1831 discovered that a change in magnetic flux induces an electromotive force in a loop of wire—a principle known as electromagnetic induction that helps explain how generators and transformers work.

In 1881 two electricians built the world's first power station at Godalming in England. The station employed two waterwheels to produce an alternating current that was used to supply seven Siemens arc lamps at 250 volts and thirty-four incandescent lamps at 40 volts. However supply was intermittent and in 1882 Thomas Edison and his company, The Edison Electric Light Company, developed the first steam-powered electric power station on Pearl Street in New York City. The Pearl Street Station consisted of several generators and initially powered around 3,000 lamps for 59 customers. The power station used direct current and operated at a single voltage. Since the direct current power could not be easily transformed to the higher voltages necessary to minimise power loss during transmission, the possible distance between the generators and load was limited to around half-a-mile (800 m).

That same year in London Lucien Gaulard and John Dixon Gibbs demonstrated the first transformer suitable for use in a real power system. The practical value of Gaulard and Gibbs› transformer was demonstrated in 1884 at Turin where the transformer was used to light up forty kilometres (25 miles) of railway from a single alternating current generator. Despite the success of the system, the pair made some fundamental mistakes. Perhaps the most serious was connecting the primaries of the transformers in series so that switching one lamp on or off would affect other lamps further down the line. Following the demonstration George Westinghouse, an American entrepreneur, imported a number of the transformers along with a Siemens generator and set his engineers to experimenting with them in the hopes of improving them for use in a commercial power system.

One of Westinghouse›s engineers, William Stanley, recognised the problem with connecting transformers in series as opposed to parallel and also realised that making the iron core of a transformer a fully enclosed loop would improve the voltage regulation of the secondary winding. Using this knowledge he built the worlds first practical transformer based alternating current power system at Great Barrington, Massachusetts in 1886. In 1885 the Italian physicist and electrical engineer Galileo Ferraris demonstrated an induction motor and in 1887 and 1888 the Serbian-American engineer Nikola Tesla filed a range of patents related to power systems including one for a practical two-phase induction motor which Westinghouse licensed for his AC system.

By 1890 the power industry had flourished and power companies had built thousands of power systems (both direct and alternating current) in the United States and Europe – these networks were effectively dedicated to providing electric lighting. During this time a fierce rivalry in the US known as the «War of Currents» emerged between Edison and Westinghouse over which form of transmission (direct or alternating current) was superior. In 1891, Westinghouse installed the first major power system that was designed to drive an electric motor and not just provide electric lighting. The installation powered a 100 horsepower (75 kW) synchronous motor at Telluride, Colorado with the motor being started by a Tesla induction motor. On the other side of the Atlantic, Oskar von Miller built a 20 kV 176 km three-phase transmission line from Lauffen am Neckar to Frankfurt am Main for the Electrical Engineering Exhibition in Frankfurt. In 1895, after a protracted decision-making process, the Adams No. 1 generating station at Niagara Falls began transmitting three-phase alternating current power to Buffalo at 11 kV. Following completion of the Niagara Falls project, new power systems increasingly chose alternating current as opposed to direct current for electrical transmission.

Although the 1880s and 1890s were seminal decades in the field, developments in power engineering continued throughout the 20th and 21st century. In 1936 the first commercial high-voltage direct current (HVDC) line using mercury-arc valves was built between Schenectady and Mechanicville, New York. HVDC had previously been achieved by installing direct current generators in series (a system known as the Thury system) although this suffered from serious reliability issues. In 1957 Siemens demonstrated the first solid-state rectifier (solid-state rectifiers are now the standard for HVDC systems) however it was not until the early 1970s that this technology was used in commercial power systems. In 1959 Westinghouse demonstrated the first circuit breaker that used SF_6 as the interrupting medium. SF_6 is a far superior dielectric to air and, in recent times, its use has been extended to produce far more compact switching equipment (known as switchgear) and transformers. Many important developments also came from extending innovations in the ICT field to the power engineering field. For example, the development of computers meant load flow studies could be run more efficiently allowing for much better planning of power systems. Advances in information technology and telecommunication also allowed for much better remote control of the power system's switchgear and generators.

BASICS OF ELECTRIC POWER

An external AC to DC power adapter used for household appliances

Electric power is the mathematical product of two quantities: current and voltage. These two quantities can vary with respect to time (AC power) or can be kept at constant levels (DC power).

Most refrigerators, air conditioners, pumps and industrial machinery use AC power whereas computers and digital equipment use DC power (the digital devices you plug into the mains typically have an internal or external power adapter to convert from AC to DC power). AC power has the advantage of being easy to transform between voltages and is able to be generated and utilised by brushless machinery. DC power remains the only practical choice in digital systems and can be more economical to transmit over long distances at very high voltages (see HVDC).

The ability to easily transform the voltage of AC power is important for two reasons: Firstly, power can be transmitted over long distances with less loss at higher voltages. So in power networks where generation is distant from the load, it is desirable to step-up the voltage of power at the generation point and then step-down the voltage near the load. Secondly, it is often more economical to install turbines that produce higher voltages than would be used by most appliances, so the ability to easily transform voltages means this mismatch between voltages can be easily managed.

Solid state devices, which are products of the semiconductor revolution, make it possible to transform DC power to different voltages, build brushless DC machines and convert between AC and DC power. Nevertheless, devices utilising solid state technology are often more expensive than their traditional counterparts, so AC power remains in widespread use.

POWER

Transmission lines transmit power across the grid.

Power Engineering deals with the generation, transmission, distribution and utilization of electricity as well as the design of a range of related devices. These include transformers, electric generators, electric motors and power electronics.

The power grid is an electrical network that connects a variety of electric generators to the users of electric power. Users purchase electricity from the grid so that they do not need to generate their own. Power engineers may work on the design and maintenance of the power grid as well as the power systems that connect to it. Such systems are called on-grid power systems and may supply the grid with additional power, draw power from the grid or do both. The grid is designed and managed using software that performs simulations of power flows.

Power engineers may also work on systems that do not connect to the grid. These systems are called off-grid power systems and may be used in preference to on-grid systems for a variety of reasons. For example, in remote locations it may be cheaper for a mine to generate its own power rather than pay for connection to the grid and in most mobile applications connection to the grid is simply not practical.

Today, most grids adopt three-phase electric power with alternating current. This choice can be partly attributed to the ease with which this type of power can be generated, transformed and used. Often (especially in the United States), the power is split before it reaches residential customers whose low-power appliances rely upon single-phase electric power. However, many larger industries and organizations still prefer to receive the three-phase power directly because it can be used to drive highly efficient electric motors such as three-phase induction motors.

Transformers play an important role in power transmission because they allow power to be converted to and from higher voltages. This is important because higher voltages suffer less power loss during transmission. This is because higher voltages allow for lower current to deliver the same amount of power, as power is the product of the two. Thus, as the voltage steps up, the current steps down. It is the current flowing through the components that result in both the losses and the subsequent heating. These losses, appearing in the form of heat, are equal to the current squared times the electrical resistance through which the current flows, so as the voltage goes up the losses are dramatically reduced.

For these reasons, electrical substations exist throughout power grids to convert power to higher voltages before transmission and to lower voltages suitable for appliances after transmission.

Components

Power engineering is a network of interconnected components which convert different forms of energy to electrical energy. Modern power engineering consists of four main subsystems: the generation subsystem, the transmission subsystem, the distribution subsystem and the utilization subsystem. In the generation subsystem, the power plant produces the electricity. The transmission subsystem transmits the electricity to the load centers. The distribution subsystem continues to transmit the power to the customers. The utilization system is concerned with the different uses of electrical energy like illumination, refrigeration, traction, electric drives, etc. Utilization is a very recent concept in Power engineering.

Generation

Generation of electrical power is a process whereby energy is transformed into an electrical form. There are several different transformation processes, among which are chemical, photo-voltaic, and electromechanical. Electromechanical energy conversion is used in converting energy from coal, petroleum, natural gas, uranium, or water flow into electrical energy. Of these, all except the wind energy conversion process take advantage of the synchronous AC generator coupled to a steam, gas or hydro turbine such that the turbine converts steam, gas, or water flow into rotational energy, and the synchronous generator then converts the rotational energy of the turbine into electrical energy. It is the turbine-generator conversion process that is by far most economical and consequently most common in the industry today.

The AC synchronous machine is the most common technology for generating electrical energy. It is called synchronous because the composite magnetic field produced by the three stator windings rotate at the same speed as the magnetic field produced by the field winding on the rotor. A simplified circuit model is used to analyze steady-state operating conditions for a synchronous machine. The phasor diagram is an effective tool for visualizing the relationships between internal voltage, armature current, and terminal voltage. The excitation control system is used on synchronous machines to regulate terminal voltage, and the turbine-governor system is used to regulate the speed of the machine. However, in highly interconnected systems, such as the "Western system", the "Texas system" and the "Eastern system", *one* machine will usually be assigned as the so-called "swing machine", and which generation may be increased or decreased to compensate for small changes in load, thereby maintaining the system frequency at precisely 60 Hz. Should the load dramatically change, as happens with a system separation, then a combination of "spinning reserve" and the "swing machine" may be used by the system's load dispatcher.

The operating costs of generating electrical energy is determined by the fuel cost and the efficiency of the power station. The efficiency depends on generation level and can be obtained from the heat rate curve. The incremental cost curve may also be derived from the heat rate curve. *Economic dispatch* is the process of allocating the required load demand between the available generation units such that the cost of operation is minimized. *Emission dispatch* is the process of allocating the required load demand between the available generation units such that air pollution occurring from operation is minimized. In large systems, particularly in the West, a combination of economic and emission dispatch may be used.

Transmission

The electricity is transported to load locations from a power station to a transmission subsystem. Therefore, we may think of the transmission system as providing the medium of transportation for electric energy. The transmission system may be subdivided into the bulk transmission system and the sub-transmission system. The functions of the bulk transmission are to interconnect generators, to interconnect various areas of the network, and to transfer electrical energy from the generators to the major load centers. This portion of the system is called "bulk" because it delivers energy only to so-called bulk loads such as the distribution system of a town, city, or large industrial plant. The function of the sub-transmission system is to interconnect the bulk power system with the distribution system.

Transmission circuits may be built either underground or overhead. Underground cables are used predominantly in urban areas where acquisition of overhead rights of way are costly or not possible. They are also used for transmission under rivers, lakes and bays. Overhead transmission is used otherwise because, for a given voltage level, overhead conductors are much less expensive than underground cables.

The transmission system is a highly integrated system. It is referred to as the substation equipment and transmission lines. The substation equipment contain the transformers, relays, and circuit breakers. Transformers are important static devices which transfer electrical energy from one circuit to another in the transmission subsystem. Transformers are used to step up the voltage on the transmission line to reduce the power loss which is dissipated on the way. A relay is functionally a level-detector; they perform a switching action when the input voltage (or current) meets or exceeds a specific and adjustable value. A circuit breaker is an automatically operated electrical switch designed to protect an electrical circuit from damage caused by overload or short circuit. A change in the status of any one component can significantly affect the operation of the entire system. Without adequate contact protection, the occurrence of undesired electric arcing causes significant degradation of the contacts, which suffer serious damage. There are three possible causes for power flow limitations to a transmission line. These causes are thermal overload, voltage instability, and rotor angle instability. Thermal overload is caused by excessive current flow in a circuit causing overheating. Voltage instability is said to occur when the power required to maintain voltages at or above acceptable levels exceeds the available power. Rotor angle instability is a dynamic problem that may occur following faults, such as short circuit, in the transmission system. It may also occur tens of seconds after a fault due to poorly damped or undamped oscillatory response of the rotor motion. As long as the *equal area criteria* is maintained, the interconnected system will remain stable. Should the *equal area criteria* be violated, it becomes necessary to separate the unstable component from the remainder of the system.

Distribution

The distribution system transports the power from the transmission system/substation to the customer. Distribution feeders can be radial or networked in an open loop configuration with a single or multiple alternate sources. Rural systems tend to be of the former and urban systems the latter. The equipment associated with the distribution system usually begins downstream of the distribution feeder circuit breaker. The transformer and circuit breaker are usually under the jurisdiction of a "substations department". The distribution feeders consist of combinations of overhead and underground conductor, 3 phase and single phase switches with load break and non-loadbreak ability, relayed protective devices, fuses, transformers (to utilization voltage), surge arresters, voltage regulators and capacitors.

More recently, Smart Grid initiatives are being deployed so that: One, distribution feeder faults are automatically isolated and power restored to unfaulted circuits by automatic hardware/software/communications packages. And two, capacitors are automatically switched on or off to dynamically control VAR flow and for CVR (Conservation Voltage Reduction).

Utilization

Utilization is the "end result" of the generation, transmission, and distribution of electric power. The energy carried by the transmission and distribution system is turned into useful work, light, heat, or a combination of these items at the utilization point. Understanding and characterizing the utilization of electric power is critical for proper planning and operation of power systems. Improper characterization of utilization can result of over or under building of power system facilities and stressing of system equipment beyond design capabilities. The term load refers to a device or collection of devices that draw energy from the power system. Individual loads (devices) range from small light bulbs to large induction motors to arc furnaces. The term load is often somewhat arbitrarily applied, at times being used to describe a specific device, and other times referring to an entire facility and even being used to describe the lumped power requirements of power system components and connected utilization devices downstream of a specific point in large scale system studies.

A major application of electric energy is in its conversion to mechanical energy. Electromagnetic, or "EM" devices designed for this purpose are commonly called "motors." Actually the machine is the central component of an integrated system consisting of the source, controller, motor, and load. For specialized applications, the system may be, and frequently is, designed as an integrated whole. Many household appliances (e.g., a vacuum cleaner) have in one unit, the controller, the motor, and the load.

REFERENCES

1. "Pioneers in Electricity and Magnetism: William Gilbert". National High Magnetic Field Laboratory. Retrieved 2008-05-25.
2. "The History Of The Light Bulb". Net Guides Publishing, Inc. 2004. Retrieved 2007-05-02.
3. Greenslade, Thomas. "The Voltaic Pile". Kenyon College. Retrieved 2008-03-31.
4. "Faraday Page". The Royal Institute. Retrieved 2008-03-31.
5. "Godalming Power Station". Engineering Timelines. Retrieved 2009-05-03.
6. Williams, Jasmin (2007-11-30). "Edison Lights The City". New York Post. Retrieved 2008-03-31.
7. Grant, Casey. "The Birth of NFPA". National Fire Protection Association. Retrieved 2008-03-31.
8. "Bulk Electricity Grid Beginnings" (PDF) (Press release). New York Independent System Operator. Retrieved 2008-05-25.
9. Katz, Evgeny (2007-04-08). "Lucien Gaulard". Archived from the original on 2008-04-22. Retrieved 2008-05-25.
10. Great Barrington 1886 - Inspiring an industry toward AC power
11. Blalock, Thomas (2004-10-02). "Alternating Current Electrification, 1886". IEEE. Retrieved 2008-05-25.
12. Fritz E. Froehlich, Allen Kent, The Froehlich/Kent Encyclopedia of Telecommunications: Volume 17, page 36. Books.google.com. Retrieved 2012-09-10.
13. Petar Miljanic, Tesla's Polyphase System and Induction Motor, Serbian Journal of Electrical Engineering, pp. 121–130, Vol. 3, No. 2, November 2006.
14. Foran, Jack. "The Day They Turned The Falls On". Retrieved 2008-05-25.
15. Voith Siemens (company) (2007-02-01). HyPower (PDF). p. 7.

16. "Adams Hydroelectric Generating Plant, 1895". IEEE. Retrieved 2008-05-25.
17. "A Novel but Short-Lived Power Distribution System". IEEE. 2005-05-01. Retrieved 2008-05-25.
18. Gene Wolf (2000-12-01). "Electricity Through the Ages". Transmission & Distribution World.
19. John Tyner, Rick Bush and Mike Eby (1999-11-01). "A Fifty-Year Retrospective". Transmission & Distribution World.
20. "Gas Insulated Switchgear". ABB. Retrieved 2008-05-25.
21. Amin, Sayed. "SF6 Transformer". Retrieved 2008-05-25.
22. All About Circuits [Online textbook], Tony R. Kuphaldt et al., last accessed on 17 May 2009.
23. Roberto Rudervall, J.P. Charpentier and Raghuveer Sharma (March 7–8, 2000). "High Voltage Direct Current (HVDC) Transmission Systems Technology Review Paper" (PDF). World Bank. (also here)
24. Ned Mohan, T. M. Undeland and William P. Robbins (2003). Power Electronics: Converters, Applications, and Design. United States of America: John Wiley & Sons, Inc. ISBN 0-471-22693-9.
25. Transformers
26. "Lab Note #105 Contact Life – Unsuppressed vs. Suppressed Arcing". Arc Suppression Technologies. April 2011. Retrieved February 5, 2012.

CHAPTER 5

Electric Motor

Chapter Contents

- **Introduction**
- **Motor Supply And Control**
- **Dc Motor**
- **Brushless Dc Electric Motor**
- **Switched Reluctance Motor**
- **Universal Motor**
- **Ac Motor**
- **Special Magnetic Motors**

The motor or an electrical motor is a device that has brought about one of the biggest advancements in the fields of engineering and technology ever since the invention of electricity. A motor is nothing but an electro-mechanical device that converts electrical energy to mechanical energy. Its because of motors, life is what it is today in the 21st century. Without motor we had still been living in Sir Thomas Edison's Era where the only purpose of electricity would have been to glow bulbs. There are different types of motor have been developed for different specific purposes.

INTRODUCTION

An electric motor is an electrical machine that converts electrical energy into mechanical energy. The reverse of this is the conversion of mechanical energy into electrical energy and is done by an electric generator.

In normal motoring mode, most electric motors operate through the interaction between an electric motor`s magnetic field and winding currents to generate force within the motor. In certain applications, such as in the transportation industry with traction motors, electric motors can operate in both motoring and generating or braking modes to also produce electrical energy from mechanical energy.

Found in applications as diverse as industrial fans, blowers and pumps, machine tools, household appliances, power tools, and disk drives, electric motors can be powered by direct current (DC) sources, such as from batteries, motor vehicles or rectifiers, or by alternating current (AC) sources, such as from the power grid, inverters or generators. Small motors may be found in electric watches. General-purpose motors with highly standardized dimensions and characteristics provide convenient mechanical power for industrial use. The largest of electric motors are used for ship propulsion, pipeline compression and pumped-storage applications with ratings reaching 100 megawatts. Electric motors may be classified by electric power source type, internal construction, application, type of motion output, and so on.

Electric motors are used to produce linear or rotary force (torque), and should be distinguished from devices such as magnetic solenoids and loudspeakers that convert electricity into motion but do not generate usable mechanical powers, which are respectively referred to as actuators and transducers.

Various electric motors, compared to 9 V battery.

Cutaway view through stator of induction motor.

Early Motors

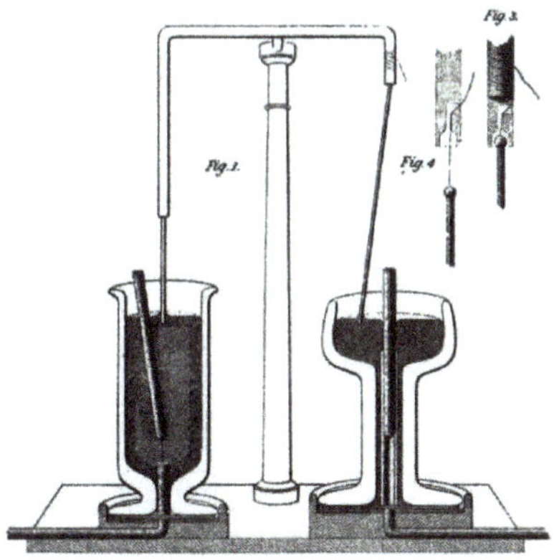

Faraday's electromagnetic experiment, 1821

Perhaps the first electric motors were simple electrostatic devices created by the Scottish monk Andrew Gordon in the 1740s. The theoretical principle behind production of mechanical force by the interactions of an electric current and a magnetic field, Ampère's force law, was discovered later by André-Marie Ampère in 1820. The conversion of electrical energy into mechanical energy by electromagnetic means was demonstrated by the British scientist Michael Faraday in 1821. A free-hanging wire was dipped into a pool of mercury, on which a permanent magnet (PM) was placed. When a current was passed through the wire, the wire rotated around the magnet, showing that the current gave rise to a close circular magnetic field around the wire. This motor is often demonstrated in physics experiments, brine substituting for toxic mercury. Though Barlow's wheel was an early refinement to this Faraday demonstration, these and similar homopolar motors were to remain unsuited to practical application until late in the century.

Jedlik's "electromagnetic self-rotor", 1827 (Museum of Applied Arts, Budapest). The historic motor still works perfectly today.

In 1827, Hungarian physicist Ányos Jedlik started experimenting with electromagnetic coils. After Jedlik solved the technical problems of the continuous rotation with the invention of the commutator, he called his early devices "electromagnetic self-rotors".

Although they were used only for instructional purposes, in 1828 Jedlik demonstrated the first device to contain the three main components of practical DC motors: the stator, rotor and commutator. The device employed no permanent magnets, as the magnetic fields of both the stationary and revolving components were produced solely by the currents flowing through their windings.

Success with DC Motors

After many other more or less successful attempts with relatively weak rotating and reciprocating apparatus the Prussian Moritz von Jacobi created the first real rotating electric motor in May 1834 that actually developed a remarkable mechanical output power. His motor set a world record which was improved only four years later in September 1838 by Jacobi himself. His second motor was powerful enough to drive a boat with 14 people across a wide river. It was not until 1839/40 that other developers worldwide managed to build motors of similar and later also of higher performance.

The first commutator DC electric motor capable of turning machinery was invented by the British scientist William Sturgeon in 1832. Following Sturgeon's work, a commutator-type direct-current electric motor made with the intention of commercial use was built by the American inventor Thomas Davenport, which he patented in 1837. The motors ran at up to 600 revolutions per minute, and powered machine tools and a printing press. Due to the high cost of primary battery power, the motors were commercially unsuccessful and Davenport went bankrupt. Several inventors followed Sturgeon in the development of DC motors but all encountered the same battery power cost issues. No electricity distribution had been developed at the time. Like Sturgeon's motor, there was no practical commercial market for these motors.

In 1855, Jedlik built a device using similar principles to those used in his electromagnetic self-rotors that was capable of useful work. He built a model electric vehicle that same year.

A major turning point in the development of DC machines took place in 1864, when Antonio Pacinotti described for the first time the ring armature with its symmetrically grouped coils closed upon themselves and connected to the bars of a commutator, the brushes of which delivered practically non-fluctuating current. The first commercially successful DC motors followed the invention by Zénobe Gramme who, in 1871, reinvented Pacinotti`s design. In 1873, Gramme showed that his dynamo could be used as a motor, which he demonstrated to great effect at exhibitions in Vienna and Philadelphia by connecting two such DC motors at a distance of up to 2 km away from each other, one as a generator.

In 1886, Frank Julian Sprague invented the first practical DC motor, a non-sparking motor that maintained relatively constant speed under variable loads. Other Sprague electric inventions about this time greatly improved grid electric distribution (prior work done while employed by Thomas Edison), allowed power from electric motors to be returned to the electric grid, provided for electric distribution to trolleys via overhead wires and the trolley pole, and provided controls systems for electric operations. This allowed Sprague to use electric motors to invent the first electric trolley system in 1887–88 in Richmond VA, the electric elevator and control system in 1892, and the electric subway with independently powered centrally controlled cars, which were first installed in 1892 in Chicago by the South Side Elevated Railway where it became popularly known as the "L". Sprague`s motor and related inventions led to an explosion of interest and use in electric motors for industry, while almost simultaneously another great inventor was developing its primary competitor, which would become much more widespread. The development of electric motors of acceptable efficiency was delayed for several decades by failure to recognize the extreme importance of a relatively small air gap between rotor and stator. Efficient designs have a comparatively small air gap. The St. Louis motor, long used in classrooms to illustrate motor principles, is extremely inefficient for the same reason, as well as appearing nothing like a modern motor.

Application of electric motors revolutionized industry. Industrial processes were no longer limited by power transmission using line shafts, belts, compressed air or hydraulic pressure. Instead every machine could be equipped with its own electric motor, providing easy control at the point of use, and improving power transmission efficiency. Electric motors applied in agriculture eliminated human and animal muscle power from such tasks as handling grain or pumping water. Household uses of electric motors reduced heavy labor in the home and made higher standards of convenience, comfort and safety possible. Today, electric motors stand for more than half of the electric energy consumption in the US.

Emergence of AC Motors

In 1824, the French physicist François Arago formulated the existence of rotating magnetic fields, termed Arago`s rotations, which, by manually turning switches on and off, Walter Baily demonstrated in 1879 as in effect the first primitive induction motor. In the 1880s, many inventors were trying to develop workable AC motors because AC`s advantages in long-distance high-voltage transmission were counterbalanced by the inability to operate motors on AC. The first alternating-current commutatorless induction motors were independently invented by Galileo Ferraris and Nikola Tesla, a working motor model having been demonstrated by the former in 1885 and by the latter in 1887. In 1888, the *Royal Academy of Science of Turin* published Ferraris`s research detailing the foundations of motor operation while however concluding that "the apparatus based on that principle could not be of any commercial importance as motor."

In 1888, Tesla presented his paper *A New System for Alternating Current Motors and Transformers* to the AIEE that described three patented two-phase four-stator-pole motor types: one with a four-pole rotor forming a non-self-starting reluctance motor, another with a wound rotor forming a self-starting induction motor, and the third a true synchronous motor with separately excited DC supply to rotor winding.

One of the patents Tesla filed in 1887, however, also described a shorted-winding-rotor induction motor. George Westinghouse promptly bought Tesla`s patents, employed Tesla to develop them, and assigned C. F. Scott to help Tesla; however, Tesla left for other pursuits in 1889. The constant speed AC induction motor was found not to be suitable for street cars, but Westinghouse engineers successfully adapted it to power a mining operation in Telluride, Colorado in 1891.

Steadfast in his promotion of three-phase development, Mikhail Dolivo-Dobrovolsky invented the three-phase cage-rotor induction motor in 1889 and the three-limb transformer in 1890. This type of motor is now used for the vast majority of commercial applications. However, he claimed that Tesla's motor was not practical because of two-phase pulsations, which prompted him to persist in his three-phase work. Although Westinghouse achieved its first practical induction motor in 1892 and developed a line of polyphase 60 hertz induction motors in 1893, these early Westinghouse motors were two-phase motors with wound rotors until B. G. Lamme developed a rotating bar winding rotor.

The General Electric Company began developing three-phase induction motors in 1891. By 1896, General Electric and Westinghouse signed a cross-licensing agreement for the bar-winding-rotor design, later called the squirrel-cage rotor. Induction motor improvements flowing from these inventions and innovations were such that a 100 horsepower (HP) induction motor currently has the same mounting dimensions as a 7.5 HP motor in 1897.

Motor Construction

Electric motor rotor (left) and stator (right)

Rotor

In an electric motor the moving part is the rotor which turns the shaft to deliver the mechanical power. The rotor usually has conductors laid into it which carry currents that interact with the magnetic field of the stator to generate the forces that turn the shaft. However, some rotors carry permanent magnets, and the stator holds the conductors.

Stator

The stator is the stationary part of the motor's electromagnetic circuit and usually consists of either windings or permanent magnets. The stator core is made up of many thin metal sheets, called laminations. Laminations are used to reduce energy losses that would result if a solid core were used.

Air Gap

The distance between the rotor and stator is called the air gap. The air gap has important effects, and is generally as small as possible, as a large gap has a strong negative effect on the performance of an electric motor. It is the main source of the low power factor at which motors operate.The air gap increases the magnetizing current needed. For this reason air gap should be minimum . Very small gaps may pose mechanical problems in addition to noise and losses.

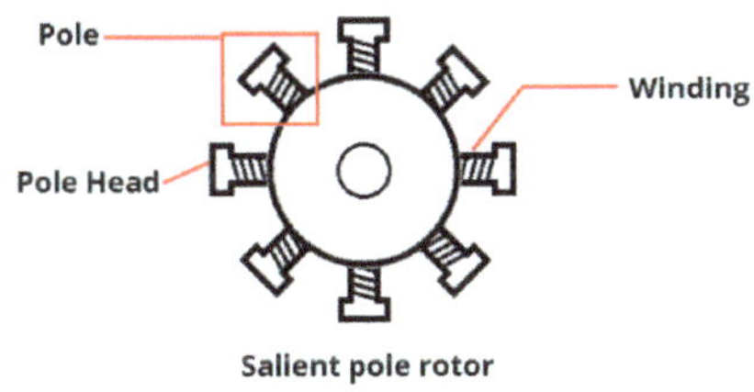

Salient-pole rotor

Windings

Windings are wires that are laid in coils, usually wrapped around a laminated soft iron magnetic core so as to form magnetic poles when energized with current.

Electric machines come in two basic magnet field pole configurations: *salient-pole* machine and *nonsalient-pole* machine. In the salient-pole machine the pole's magnetic field is produced by a winding wound around the pole below the pole face. In the *nonsalient-pole*, or distributed field, or round-rotor, machine, the winding is distributed in pole face slots. A shaded-pole motor has a winding around part of the pole that delays the phase of the magnetic field for that pole.

Some motors have conductors which consist of thicker metal, such as bars or sheets of metal, usually copper, although sometimes aluminum is used. These are usually powered by electromagnetic induction.

Commutator

A toy's small DC motor with its commutator

A commutator is a mechanism used to switch the input of most DC machines and certain AC machines consisting of slip ring segments insulated from each other and from the electric motor's shaft. The motor's armature current is supplied through the stationary brushes in contact with the revolving commutator, which causes required current reversal and applies power to the machine in an optimal manner as the rotor rotates from pole to pole. In absence of such current reversal, the motor would brake to a stop. In light of significant advances in the past few decades due to improved technologies in electronic controller, sensorless control, induction motor, and permanent magnet motor fields, electromechanically commutated motors are increasingly being displaced by externally commutated induction and permanent-magnet motors.

MOTOR SUPPLY AND CONTROL

Motor Supply

A DC motor is usually supplied through slip ring commutator as described above. AC motors' commutation can be either slip ring commutator or externally commutated type, can be fixed-speed or variable-speed control type, and can be synchronous or asynchronous type. Universal motors can run on either AC or DC.

Motor Control

Fixed-speed controlled AC motors are provided with direct-on-line or soft-start starters.

Variable speed controlled AC motors are provided with a range of different power inverter, variable-frequency drive or electronic commutator technologies.

The term electronic commutator is usually associated with self-commutated brushless DC motor and switched reluctance motor applications.

Major Categories

Electric motors operate on three different physical principles: magnetic, electrostatic and piezoelectric. By far the most common is magnetic.

In magnetic motors, magnetic fields are formed in both the rotor and the stator. The product between these two fields gives rise to a force, and thus a torque on the motor shaft. One, or both, of these fields must be made to change with the rotation of the motor. This is done by switching the poles on and off at the right time, or varying the strength of the pole.

The main types are DC motors and AC motors, the former increasingly being displaced by the latter.

AC electric motors are either asynchronous or synchronous.

Once started, a synchronous motor requires synchronism with the moving magnetic field's synchronous speed for all normal torque conditions.

In synchronous machines, the magnetic field must be provided by means other than induction such as from separately excited windings or permanent magnets.

A fractional horsepower (FHP) motor either has a rating below about 1 horsepower (0.746 kW), or is manufactured with a standard frame size smaller than a standard 1 HP motor. Many household and industrial motors are in the fractional horsepower class.

Type of Motor Commutation				
Major Categories by Self-Commutated			**Externally Commutated**	
Mechanical-Commutator Motors		**Electronic-Commutator (EC) Motors**	**Asynchronous Machines**	**Synchronous Machines**[2]
AC	**DC**	**AC**[5, 6]	**AC**[6]	
*** Universal motor (AC commutator series motor or AC/DC motor)**[1] *** Repulsion motor**	**Electrically excited DC motor:** *** Separately excited** *** Series** *** Shunt** *** Compound** **PM DC motor**	**With PM rotor:** *** BLDC motor** **With ferromagnetic rotor:** *** SRM**	**Three-phase motors:** *** SCIM**[3, 8] *** WRIM**[4, 7, 8] **AC motors:**[10] *** Capacitor** *** Resistance** *** Split** *** Shaded-pole**	**Three-phase motors:** *** WRSM** *** PMSM or BLAC motor** **- IPMSM** **- SPMSM** *** Hybrid** **AC motors:**[10] *** Permanent-split capacitor** *** Hysteresis** *** Stepper** *** SyRM** *** SyRM-PM hybrid**

Notes:

- Rotation is independent of the frequency of the AC voltage.
- Rotation is equal to synchronous speed (motor stator field speed).
- In SCIM fixed-speed operation rotation is equal to synchronous speed less slip speed.
- In non-slip energy recovery systems WRIM is usually used for motor starting but can be used to vary load speed.
- Variable-speed operation.
- Whereas induction and synchronous motor drives are typically with either six-step or sinusoidal waveform output, BLDC motor drives are usually with trapezoidal current waveform; the behavior of both sinusoidal and trapezoidal PM machines is however identical in terms of their fundamental aspects.
- In variable-speed operation WRIM is used in slip energy recovery and double-fed induction machine applications.

- A cage winding is a shorted-circuited squirrel-cage rotor, a wound winding is connected externally through slip rings.
- Mostly single-phase with some three-phase.

Abbreviations:

- BLAC - Brushless AC
- BLDC - Brushless DC
- BLDM - Brushless DC motor
- EC - Electronic commutator
- PM - Permanent magnet
- IPMSM - Interior permanent magnet synchronous motor
- PMSM - Permanent magnet synchronous motor
- SPMSM - Surface permanent magnet synchronous motor
- SCIM - Squirrel-cage induction motor
- SRM - Switched reluctance motor
- SyRM - Synchronous reluctance motor
- VFD - Variable-frequency drive
- WRIM - Wound-rotor induction motor
- WRSM - Wound-rotor synchronous motor

Self-Commutated Motor

Brushed DC Motor

All self-commutated DC motors are by definition run on DC electric power. Most DC motors are small PM types. They contain a brushed internal mechanical commutation to reverse motor windings' current in synchronism with rotation.

DC MOTOR

A DC motor is any of a class of rotary electrical machines that converts direct current electrical power into mechanical power. The most common types rely on the forces produced by magnetic fields. Nearly all types of DC motors have some internal mechanism, either electromechanical or electronic, to periodically change the direction of current flow in part of the motor.

DC motors were the first type widely used, since they could be powered from existing direct-current lighting power distribution systems. A DC motor's speed can be controlled over a wide range, using either a variable supply voltage or by changing the strength of current in its field windings. Small DC motors are used in tools, toys, and appliances. The universal motor can operate on direct current but is a lightweight motor used for portable power tools and appliances. Larger DC motors are used in propulsion of electric vehicles, elevator and hoists, or in drives for steel rolling mills. The advent of power electronics has made replacement of DC motors with AC motors possible in many applications.

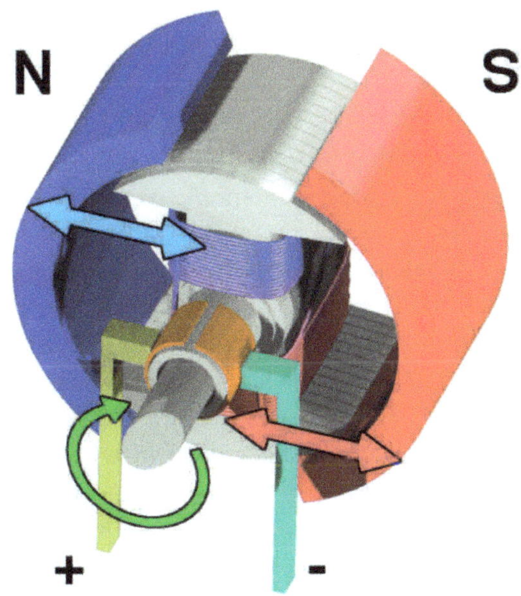

Workings of a brushed electric motor with a two-pole rotor (armature) and permanent magnet stator. "N" and "S" designate polarities on the inside axis faces of the magnets; the outside faces have opposite polarities. The + and - signs show where the DC current is applied to the commutator which supplies current to the armature coils

Electromagnetic Motors

A coil of wire with a current running through it generates an electromagnetic field aligned with the center of the coil. The direction and magnitude of the magnetic field produced by the coil can be changed with the direction and magnitude of the current flowing through it.

A simple DC motor has a stationary set of magnets in the stator and an armature with one or more windings of insulated wire wrapped around a soft iron core that concentrates the magnetic field. The windings usually have multiple turns around the core, and in large motors there can be several parallel current paths. The ends of the wire winding are connected to a commutator. The commutator allows each armature coil to be energized in turn and connects the rotating coils with the external power supply through brushes. (Brushless DC motors have electronics that switch the DC current to each coil on and off and have no brushes.)

The total amount of current sent to the coil, the coil's size and what it's wrapped around dictate the strength of the electromagnetic field created.

The sequence of turning a particular coil on or off dictates what direction the effective electromagnetic fields are pointed. By turning on and off coils in sequence a rotating magnetic field can be created. These rotating magnetic fields interact with the magnetic fields of the magnets (permanent or electromagnets) in the stationary part of the motor (stator) to create a force on the armature which causes it to rotate. In some DC motor designs the stator fields use electromagnets to create their magnetic fields which allow greater control over the motor. At high power levels, DC motors are almost always cooled using forced air.

Different number of stator and armature fields as well as how they are connected provide different inherent speed/torque regulation characteristics. The speed of a DC motor can be controlled by changing the voltage applied to the armature. The introduction of variable resistance in the armature circuit or field circuit allowed speed control. Modern DC motors are often controlled by power electronics systems which adjust the voltage by "chopping" the DC current into on and off cycles which have an effective lower voltage.

Since the series-wound DC motor develops its highest torque at low speed, it is often used in traction applications such as electric locomotives, and trams. The DC motor was the mainstay of electric traction drives on both electric and diesel-electric locomotives, street-cars/trams and diesel electric drilling rigs for many years. The introduction of DC motors and an electrical grid system to run machinery starting in the 1870s started a new second Industrial Revolution. DC motors can operate directly from rechargeable batteries, providing the motive power for the first electric vehicles and today's hybrid cars and electric cars as well as driving a host of cordless tools. Today DC motors are still found in applications as small as toys and disk drives, or in large sizes to operate steel rolling mills and paper machines. Large DC motors with separately excited fields were generally used with winder drives for mine hoists, for high torque as well as smooth speed control using thyristor drives. These are now replaced with large AC motors with variable frequency drives.

If external power is applied to a DC motor it acts as a DC generator, a dynamo. This feature is used to slow down and recharge batteries on hybrid car and electric cars or to return electricity back to the electric grid used on a street car or electric powered train line when they slow down. This process is called regenerative braking on hybrid and electric cars. In diesel electric locomotives they also use their DC motors as generators to slow down but dissipate the energy in resistor stacks. Newer designs are adding large battery packs to recapture some of this energy.

Brushed

A brushed DC electric motor generating torque from DC power supply by using an internal mechanical commutation. Stationary permanent magnets form the stator field. Torque is produced by the principle that any current-carrying conductor placed within an external magnetic field experiences a force, known as Lorentz force. In a motor, the magnitude of this Lorentz force (a vector represented by the green arrow), and thus the output torque,is a function for rotor angle, leading to a phenomenon known as torque ripple) Since this is a single phase two-pole motor, the commutator consists of a split ring, so that the current reverses each half turn (180 degrees).

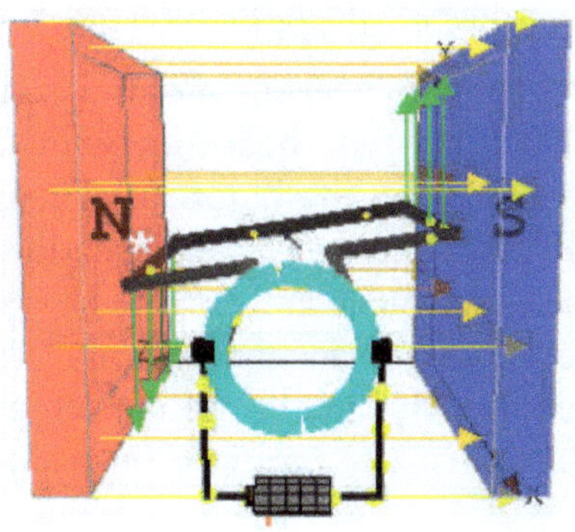

The brushed DC electric motor generates torque directly from DC power supplied to the motor by using internal commutation, stationary magnets (permanent or electromagnets), and rotating electrical magnets.

Advantages of a brushed DC motor include low initial cost, high reliability, and simple control of motor speed. Disadvantages are high maintenance and low life-span for high intensity uses. Maintenance involves regularly replacing the carbon brushes and springs which carry the electric current, as well as cleaning or replacing the commutator. These components are necessary for transferring electrical power from outside the motor to the spinning wire windings of the rotor inside the motor. Brushes consist of conductors.

Brushless

Typical brushless DC motors use one or more permanent magnets in the rotor and electromagnets on the motor housing for the stator. A motor controller converts DC to AC. This design is mechanically simpler than that of brushed motors because it eliminates the complication of transferring power from outside the motor to the spinning rotor. The motor controller can sense the rotor's position via Hall effect sensors or similar devices and can precisely control the timing, phase, etc., of the current in the rotor coils to optimize torque, conserve power, regulate speed, and even apply some braking. Advantages of brushless motors include long life span, little or no maintenance, and high efficiency. Disadvantages include high initial cost, and more complicated motor speed controllers. Some such brushless motors are sometimes referred to as "synchronous motors" although they have no external power supply to be synchronized with, as would be the case with normal AC synchronous motors.

Uncommutated

Other types of DC motors require no commutation.

- Homopolar motor – A homopolar motor has a magnetic field along the axis of rotation and an electric current that at some point is not parallel to the magnetic field. The name homopolar refers to the absence of polarity change.

Homopolar motors necessarily have a single-turn coil, which limits them to very low voltages. This has restricted the practical application of this type of motor.

- Ball bearing motor – A ball bearing motor is an unusual electric motor that consists of two ball bearing-type bearings, with the inner races mounted on a common conductive shaft, and the outer races connected to a high current, low voltage power supply. An alternative construction fits the outer races inside a metal tube, while the inner races are mounted on a shaft with a non-conductive section (e.g. two sleeves on an insulating rod). This method has the advantage that the tube will act as a flywheel. The direction of rotation is determined by the initial spin which is usually required to get it going.

Permanent Magnet Stators

A PM motor does not have a field winding on the stator frame, instead relying on PMs to provide the magnetic field against which the rotor field interacts to produce torque. Compensating windings in series with the armature may be used on large motors to improve commutation under load. Because this field is fixed, it cannot be adjusted for speed control. PM fields (stators) are convenient in miniature motors to eliminate the power consumption of the field winding. Most larger DC motors are of the "dynamo" type, which have stator windings. Historically, PMs could not be made to retain high flux if they were disassembled; field windings were more practical to obtain the needed amount of flux. However, large PMs are costly, as well as dangerous and difficult to assemble; this favors wound fields for large machines.

To minimize overall weight and size, miniature PM motors may use high energy magnets made with neodymium or other strategic elements; most such are neodymium-iron-boron alloy. With their higher flux density, electric machines with high-energy PMs are at least competitive with all optimally designed singly fed synchronous and induction electric machines. Miniature motors resemble the structure in the illustration, except that they have at least three rotor poles (to ensure starting, regardless of rotor position) and their outer housing is a steel tube that magnetically links the exteriors of the curved field magnets.

Wound Stators

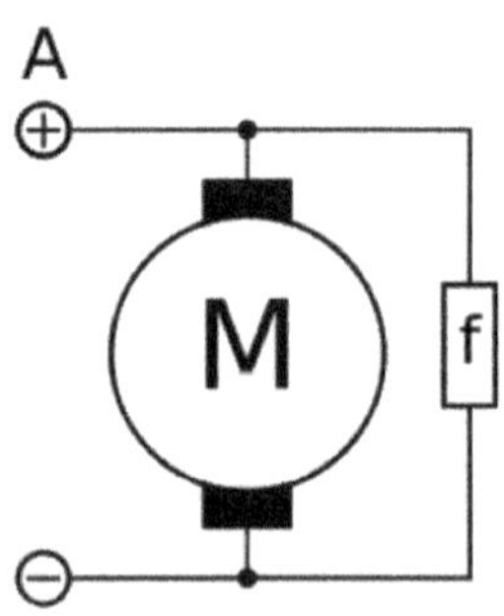

B

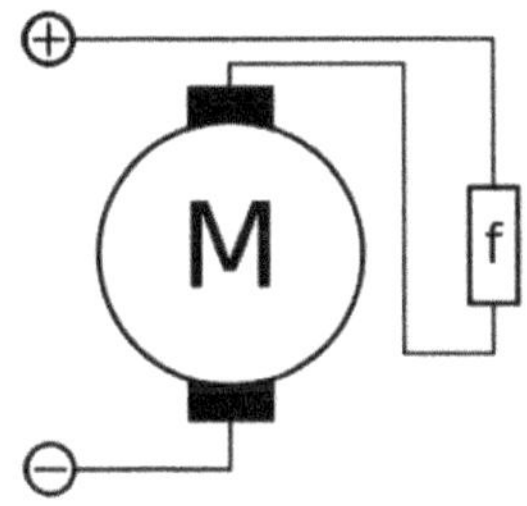

C

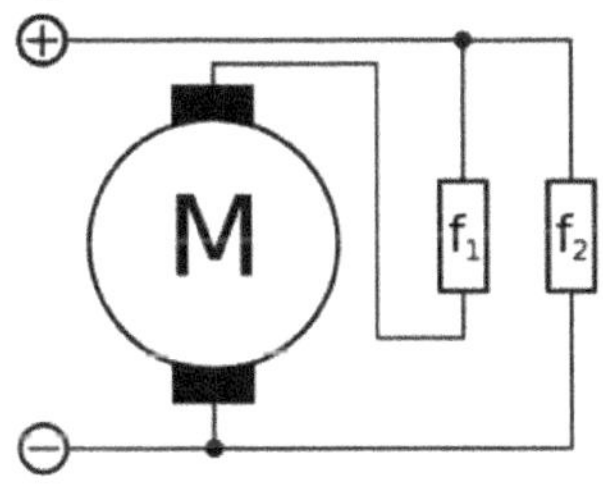

A field coil may be connected in shunt, in series, or in compound with the armature of a DC machine (motor or generator)

There are three types of electrical connections between the stator and rotor possible for DC electric motors: series, shunt/parallel and compound (various blends of series and shunt/parallel) and each has unique speed/torque characteristics appropriate for different loading torque profiles/signatures.

Series Connection

A series DC motor connects the armature and field windings in series with a common D.C. power source. The motor speed varies as a non-linear function of load torque and armature current; current is common to both the stator and rotor yielding current squared (I^2) behavior. A series motor has very high starting torque and is commonly used for starting high inertia loads, such as trains, elevators or hoists. This speed/torque characteristic is useful in applications such as dragline excavators, where the digging tool moves rapidly when unloaded but slowly when carrying a heavy load.

A series motor should never be started at no load. With no mechanical load on the series motor, the current is low, the counter-EMF produced by the field winding is weak, and so the armature must turn faster to produce sufficient counter-EMF to balance the supply voltage. The motor can be damaged by overspeed. This is called a runaway condition.

Series motors called universal motors can be used on alternating current. Since the armature voltage and the field direction reverse at the same time, torque continues to be produced in the same direction. However they run at a lower speed with lower torque on AC supply when compared to DC due to reactance voltage drop in AC which is not present in DC.Since the speed is not related to the line frequency, universal motors can develop higher-than-synchronous speeds, making them lighter than induction motors of the same rated mechanical output. This is a valuable characteristic for hand-held power tools. Universal motors for commercial utility are usually of small capacity, not more than about 1 kW output. However, much larger universal motors were used for electric locomotives, fed by special low-frequency traction power networks to avoid problems with commutation under heavy and varying loads.

Shunt Connection

A shunt DC motor connects the armature and field windings in parallel or shunt with a common D.C. power source. This type of motor has good speed regulation even as the load varies, but does not have the starting torque of a series DC motor. It is typically used for industrial, adjustable speed applications, such as machine tools, winding/unwinding machines and tensioners.

Compound Connection

A compound DC motor connects the armature and fields windings in a shunt and a series combination to give it characteristics of both a shunt and a series DC motor. This motor is used when both a high starting torque and good speed regulation is needed. The motor can be connected in two arrangements: cumulatively or differentially. Cumulative compound motors connect the series field to aid the shunt field, which provides higher starting torque but less speed regulation. Differential compound DC motors have good speed regulation and are typically operated at constant speed.

Permanent Magnet DC Motor

A PM motor does not have a field winding on the stator frame, instead relying on PMs to provide the magnetic field against which the rotor field interacts to produce torque. Compensating windings in series with the armature may be used on large motors to improve commutation under load. Because this field is fixed, it cannot be adjusted for speed control. PM fields (stators) are convenient in miniature motors to eliminate the power consumption of the field winding. Most larger DC motors are of the "dynamo" type, which have stator windings. Historically, PMs could not be made to retain high flux if they were disassembled; field windings were more practical to obtain the needed amount of flux. However, large PMs are costly, as well as dangerous and difficult to assemble; this favors wound fields for large machines.

To minimize overall weight and size, miniature PM motors may use high energy magnets made with neodymium or other strategic elements; most such are neodymium-iron-boron alloy. With their higher flux density, electric machines with high-energy PMs are at least competitive with all optimally designed singly-fed synchronous and induction electric machines. Miniature motors resemble the structure in the illustration, except that they have at least three rotor poles (to ensure starting, regardless of rotor position) and their outer housing is a steel tube that magnetically links the exteriors of the curved field magnets.

Electronic Commutator (EC) Motor

Brushless DC Motor

Some of the problems of the brushed DC motor are eliminated in the BLDC design. In this motor, the mechanical "rotating switch" or commutator is replaced by an external electronic switch synchronised to the rotor's position. BLDC motors are typically 85–90% efficient or more. Efficiency for a BLDC motor of up to 96.5% have been reported, whereas DC motors with brushgear are typically 75–80% efficient.

The BLDC motor's characteristic trapezoidal back-emf waveform is derived partly from the stator windings being evenly distributed, and partly from the placement of the rotor's PMs. Also known as electronically commutated DC or inside out DC motors, the stator windings of trapezoidal BLDC motors can be with single-phase, two-phase or three-phase and use Hall effect sensors mounted on their windings for rotor position sensing and low cost closed-loop control of the electronic commutator.

BLDC motors are commonly used where precise speed control is necessary, as in computer disk drives or in video cassette recorders, the spindles within CD, CD-ROM (etc.) drives, and mechanisms within office products such as fans, laser printers and photocopiers. They have several advantages over conventional motors:

- Compared to AC fans using shaded-pole motors, they are very efficient, running much cooler than the equivalent AC motors. This cool operation leads to much-improved life of the fan's bearings.
- Without a commutator to wear out, the life of a BLDC motor can be significantly longer compared to a DC motor using brushes and a commutator. Commutation also tends to cause a great deal of electrical and RF noise; without a commutator or brushes, a BLDC motor may be used in electrically sensitive devices like audio equipment or computers.
- The same Hall effect sensors that provide the commutation can also provide a convenient tachometer signal for closed-loop control (servo-controlled) applications. In fans, the tachometer signal can be used to derive a "fan OK" signal as well as provide running speed feedback.
- The motor can be easily synchronized to an internal or external clock, leading to precise speed control.
- BLDC motors have no chance of sparking, unlike brushed motors, making them better suited to environments with volatile chemicals and fuels. Also, sparking generates ozone which can accumulate in poorly ventilated buildings risking harm to occupants' health.
- BLDC motors are usually used in small equipment such as computers and are generally used in fans to get rid of unwanted heat.
- They are also acoustically very quiet motors which is an advantage if being used in equipment that is affected by vibrations.

Modern BLDC motors range in power from a fraction of a watt to many kilowatts. Larger BLDC motors up to about 100 kW rating are used in electric vehicles. They also find significant use in high-performance electric model aircraft.

BRUSHLESS DC ELECTRIC MOTOR

Brushless DC electric motor (BLDC motors, BL motors) also known as electronically commutated motors (ECMs, EC motors) are synchronous motors that are powered by a DC electric source via an integrated inverter/switching power supply, which produces an AC electric signal to drive the motor. In this context, AC, alternating current, does not imply a sinusoidal waveform, but rather a bi-directional current with no restriction on waveform. Additional sensors and electronics control the inverter output amplitude and waveform (and therefore percent of DC bus usage/efficiency) and frequency (i.e. rotor speed).

The rotor part of a brushless motor is often a permanent magnet synchronous motor, but can also be a switched reluctance motor, or induction motor.

Brushless motors may be described as stepper motors; however, the term "stepper motor" tends to be used for motors that are designed specifically to be operated in a mode where they are frequently stopped with the rotor in a defined angular position. This page describes more general brushless motor principles, though there is overlap.

Two key performance parameters of brushless DC motors are the motor constants Kv and Km.

The motor from a 3.5" floppy disk drive. The coils, arranged radially, are made from copper wire coated with blue insulation. The rotor (upper right) has been removed and turned upside-down. The grey ring inside its cup is a permanent magnet.

Brushless vs. Brushed Motors

Brushed DC motors have been around since the mid-19th century, but brushless motors are a fairly recent arrival; a first step in the 1960s thanks to advances in solid state technology, with further improvements in the 1980s thanks to better permanent magnet materials.

Brushed DC motors develop a maximum torque when stationary, linearly decreasing as velocity increases. Some limitations of brushed motors can be overcome by brushless motors; they include higher efficiency and a lower susceptibility to mechanical wear. These benefits come at the cost of potentially less rugged, more complex, and more expensive control electronics.

A typical brushless motor has permanent magnets which rotate around a fixed armature, eliminating problems associated with connecting current to the moving armature. An electronic controller replaces the brush/commutator assembly of the brushed DC motor, which continually switches the phase to the windings to keep the motor turning. The controller performs similar timed power distribution by using a solid-state circuit rather than the brush/commutator system.

Brushless motors offer several advantages over brushed DC motors, including high torque to weight ratio, more torque per watt (increased efficiency), increased reliability, reduced noise, longer lifetime (no brush and commutator erosion), elimination of ionizing sparks from the commutator, and overall reduction of electromagnetic interference (EMI). With no windings on the rotor, they are not subjected to centrifugal forces, and because the windings are supported by the housing, they can be cooled by conduction, requiring no airflow inside the motor for cooling. This in turn means that the motor's internals can be entirely enclosed and protected from dirt or other foreign matter.

Brushless motor commutation can be implemented in software using a microcontroller or microprocessor computer, or may alternatively be implemented in analogue hardware, or in digital firmware using an FPGA. Commutation with electronics instead of brushes allows for greater flexibility and capabilities not available with brushed DC motors, including speed limiting, "micro stepped" operation for slow and/or fine motion control, and a holding torque when stationary. Controller software can be customized to the specific motor being used in the application, resulting in greater commutation efficiency.

The maximum power that can be applied to a brushless motor is limited almost exclusively by heat; too much heat weakens the magnets and may damage the winding's insulation.

When converting electricity into mechanical power, brushless motors are more efficient than brushed motors. This improvement is largely due to the frequency at which the electricity is switched determined by the position sensor feedback. Additional gains are due to the absence of brushes, which reduces mechanical energy loss due to friction. The enhanced efficiency is greatest in the no-load and low-load region of the motor's performance curve. Under high mechanical loads, brushless motors and high quality brushed motors are comparable in efficiency.

Environments and requirements in which manufacturers use brushless-type DC motors include maintenance-free operation, high speeds, and operation where sparking is hazardous (i.e. explosive environments) or could affect electronically sensitive equipment.

Controller Implementations

Because the controller implements the traditional brushes functionality it needs the rotor's orientation/position (relative to the stator coils). This is automatic in a brushed motor due to the fixed geometry of rotor shaft and brushes. Some designs use Hall effect sensors or a rotary encoder to directly measure the rotor's position. Others measure the back-EMF in the undriven coils to infer the rotor position, eliminating the need for separate Hall effect sensors, and therefore are often called *sensorless* controllers.

In a brushless DC motor, two coils are energized at a time with equal and opposite polarities: one pushes the rotor away from it while the other attracts the rotor towards it. This increases the overall torque capacity of the motor and Hall effect sensors or a rotary encoder determine which two coils have to be energized to achieve this strategy.

A typical controller contains 3 bi-directional outputs (i.e., frequency controlled three phase output), which are controlled by a logic circuit. Simple controllers employ comparators to determine when the output phase should be advanced, while more advanced controllers employ a microcontroller to manage acceleration, control speed and fine-tune efficiency.

Controllers that sense rotor position based on back-EMF have extra challenges in initiating motion because no back-EMF is produced when the rotor is stationary. This is usually accomplished by beginning rotation from an arbitrary phase, and then skipping to the correct phase if it is found to be wrong. This can cause the motor to run briefly backwards, adding even more complexity to the startup sequence. Other sensorless controllers are capable of measuring winding saturation caused by the position of the magnets to infer the rotor position.

Variations in Construction

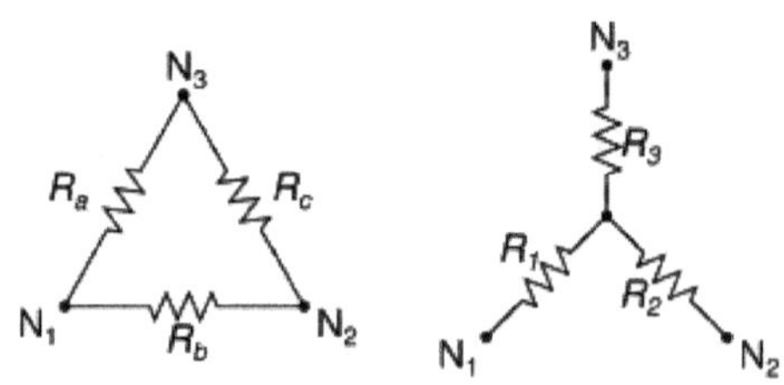

Schematic for delta and wye winding styles. (This image does not illustrate the motor's inductive and generator-like properties)

Brushless motors can be constructed in several different physical configurations: In the 'conventional' (also known as *inrunner*) configuration, the permanent magnets are part of the rotor. Three stator windings surround the rotor. In the *outrunner* (or external-rotor) configuration, the radial-relationship between the coils and magnets is reversed; the stator coils form the center (core) of the motor, while the permanent magnets spin within an overhanging rotor which surrounds the core. The flat or axial flux type, used where there are space or shape limitations, uses stator and rotor plates, mounted face to face. Outrunners typically have more poles, set up in triplets to maintain the three groups of windings, and have a higher torque at low RPMs. In all brushless motors, the coils are stationary.

There are two common electrical winding configurations; the delta configuration connects three windings to each other (series circuits) in a triangle-like circuit, and power is applied at each of the connections. The Wye (*Y*-shaped) configuration, sometimes called a star winding, connects all of the windings to a central point (parallel circuits) and power is applied to the remaining end of each winding.

A motor with windings in delta configuration gives low torque at low speed, but can give higher top speed. Wye configuration gives high torque at low speed, but not as high top speed.

Although efficiency is greatly affected by the motor's construction, the Wye winding is normally more efficient. In delta-connected windings, half voltage is applied across the windings adjacent to the driven lead (compared to the winding directly between the driven leads), increasing resistive losses. In addition, windings can allow high-frequency parasitic electrical currents to circulate entirely within the motor. A Wye-connected winding does not contain a closed loop in which parasitic currents can flow, preventing such losses.

From a controller standpoint, the two styles of windings are treated exactly the same.

Applications

The four poles on the stator of a two-phase brushless motor. This is part of a computer cooling fan; the rotor has been removed.

Brushless motors fulfill many functions originally performed by brushed DC motors, but cost and control complexity prevents brushless motors from replacing brushed motors completely in the lowest-cost areas. Nevertheless, brushless motors have come to dominate many applications, particularly devices such as computer hard drives and CD/DVD players. Small cooling fans in electronic equipment are powered exclusively by brushless motors. They can be found in cordless power tools where the increased efficiency of the motor leads to longer periods of use before the battery needs to be charged. Low speed, low power brushless motors are used in direct-drive turntables for gramophone records.

Transport

High power brushless motors are found in electric vehicles and hybrid vehicles. These motors are essentially AC synchronous motors with permanent magnet rotors.

The Segway Scooter and Vectrix Maxi-Scooter use brushless motors.

A number of electric bicycles use brushless motors that are sometimes built into the wheel hub itself, with the stator fixed solidly to the axle and the magnets attached to and rotating with the wheel.

Heating and Ventilation

There is a trend in the HVAC and refrigeration industries to use brushless motors instead of various types of AC motors. The most significant reason to switch to a brushless motor is the dramatic reduction in power required to operate them versus a typical AC motor. While shaded-pole and permanent split capacitor motors once dominated as the motor of fan, many fans are now run using a brushless motor. Some fans use brushless motors also in order to increase overall system efficiency.

In addition to the brushless motor's higher efficiency, HVAC systems (especially those featuring variable-speed and/or load modulation) use brushless motors because the built-in microprocessor allows for programmability, control over airflow, and serial communication. Some ceiling fan and portable fans are also featuring this motor. They advertise the motor being highly energy efficient and quieter than most fans.

Industrial Engineering

The application of brushless DC motors within industrial engineering primarily focuses on manufacturing engineering or industrial automation design. In manufacturing, brushless motors are primarily used for motion control, positioning or actuation systems.

Brushless motors are ideally suited for manufacturing applications because of their high power density, good speed-torque characteristics, high efficiency, wide speed ranges and low maintenance. The most common uses of brushless DC motors in industrial engineering are linear motors, servomotors, actuators for industrial robots, extruder drive motors and feed drives for CNC machine tools.

Motion Control Systems

Brushless motors are commonly used as pump, fan and spindle drives in adjustable or variable speed applications. They can develop high torque with good speed response. In addition, they can be easily automated for remote control. Due to their construction, they have good thermal characteristics and high energy efficiency. To obtain a variable speed response, brushless motors operate in an electromechanical system that includes an electronic motor controller and a rotor position feedback sensor.

Brushless DC motors are widely used as servomotors for machine tool servo drives. Servomotors are used for mechanical displacement, positioning or precision motion control. In the past DC stepper motors were used as servomotors; however, since they are operated with open loop control, they typically exhibit torque pulsations. Brushless dc motors are more suitable as servomotors since their precise motion is based upon a closed loop control system that provides tightly controlled and stable operation.

Positioning and Actuation Systems

Brushless motors are used in industrial positioning and actuation applications. For assembly robots, brushless stepper or servo motors are used to position a part for assembly or a tool for a manufacturing process, such as welding or painting. Brushless motors can also be used to drive linear actuators

Motors that directly produce linear motion are called linear motors. The advantage of linear motors is that they can produce linear motion without the need of a transmission system, such as a ball-and-lead screw, rack-and-pinion, cam, gears or belts, that would be necessary for rotary motors. Transmission systems are known to introduce less responsiveness and reduced accuracy. Direct drive, brushless DC linear motors consist of a slotted stator with magnetic teeth and a moving actuator, which has permanent magnets and coil windings. To obtain linear motion, a motor controller excites the coil windings in the actuator causing an interaction of the magnetic fields resulting in linear motion. Tubular linear motors are another form of linear motor design operated in a similar way.

Aeromodeling

A microprocessor-controlled BLDC motor powering a micro radio-controlled airplane. This external rotor motor weighs 5 g and consumes approximately 11 W.

Brushless motors are a popular motor choice for model aircraft including helicopters and drones. Their favorable power-to-weight ratios and large range of available sizes, from under 5 gram to large motors rated at well into the kilowatt output range, have revolutionized the market for electric-powered model flight, displacing virtually all brushed electric motors. They have also encouraged a growth of simple, lightweight electric model aircraft, rather than the previous internal combustion engines powering larger and heavier models. The large power-to-weight ratio of modern batteries and brushless motors allows models to ascend vertically, rather than climb gradually. The low noise and lack of mass compared to small glow fuel internal combustion engines is another reason for their popularity.

Legal restrictions for the use of combustion engine driven model aircraft in some countries, most often due to potential for noise pollution — even with purpose-designed mufflers for almost all model engines being available over the most recent decades — have also supported the shift to high-power electric systems.

Radio-Controlled Cars

Their popularity has also risen in the radio-controlled (RC) car area. Brushless motors have been legal in North American RC car racing in accordance to ROAR since 2006. These motors provide a great amount of power to RC racers and, if paired with appropriate gearing and high-discharge Li-Po (lithium polymer) or safer (but heavier and less powerful) LiFePO4 batteries, these cars can achieve speeds over 100 miles per hour (160 km/h).

Brushless motors are capable of producing more torque and have a faster peak rotational speed compared to nitro- or gasoline-powered engines. Nitro engines climax to peak at around 46,800 RPM and 2.95HP, while a smaller brushless motor has instantaneous maximum torque at start, and, when tapering off, can reach 50,000 RPM and 5HP. Larger brushless RC motors can reach upwards of 14 hp and 28,000 RPM to power 1/5th scale models.

Switched Reluctance Motor

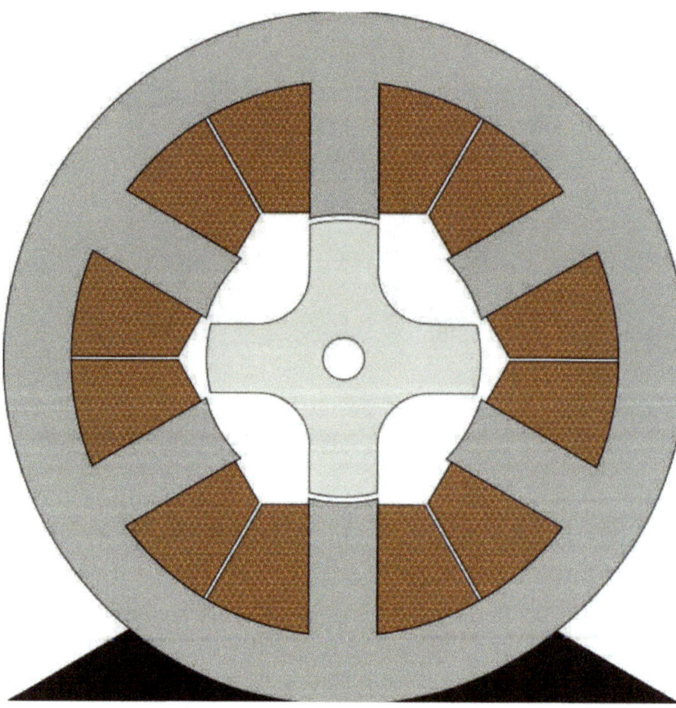

6/4 pole switched reluctance motor

The SRM has no brushes or PMs, and the rotor has no electric currents. Instead, torque comes from a slight misalignment of poles on the rotor with poles on the stator. The rotor aligns itself with the magnetic field of the stator, while the stator field windings are sequentially energized to rotate the stator field.

The magnetic flux created by the field windings follows the path of least magnetic reluctance, meaning the flux will flow through poles of the rotor that are closest to the energized poles of the stator, thereby magnetizing those poles of the rotor and creating torque. As the rotor turns, different windings will be energized, keeping the rotor turning.

SRMs are now being used in some appliances.

SWITCHED RELUCTANCE MOTOR

The switched reluctance motor (SRM) is a type of a stepper motor, an electric motor that runs by reluctance torque. Unlike common DC motor types, power is delivered to windings in the stator (case) rather than the rotor. This greatly simplifies mechanical design as power does not have to be delivered to a moving part, but it complicates the electrical design as some sort of switching system needs to be used to deliver power to the different windings. With modern electronic devices, precisely timed switching is not a problem, and the SRM is a popular design for modern stepper motors. Its main drawback is torque ripple.

An alternate use of the same mechanical design is as a generator when driven mechanically, and the load is switched to the coils in sequence to synchronize the current flow with the rotation. Such generators can be run at much higher speeds than conventional types as the armature can be made as one piece of magnetisable material, a simple slotted cylinder. In this case use of the abbreviation SRM is extended to mean Switched Reluctance Machine, although SRG, Switched Reluctance Generator is also used. A topology that is both motor and generator is useful for starting the prime mover, as it saves a dedicated starter motor.

Operating Principle

The SRM has wound field coils as in a DC motor for the stator windings. The rotor however has no magnets or coils attached. It is a solid salient-pole rotor (having projecting magnetic poles) made of soft magnetic material (often laminated-steel). When power is applied to the stator windings, the rotor's magnetic reluctance creates a force that attempts to align the rotor pole with the nearest stator pole. In order to maintain rotation, an electronic control system switches on the windings of successive stator poles in sequence so that the magnetic field of the stator "leads" the rotor pole, pulling it forward. Rather than using a troublesome high-maintenance mechanical commutator to switch the winding current as in traditional motors, the switched-reluctance motor uses an electronic position sensor to determine the angle of the rotor shaft and solid state electronics to switch the stator windings, which also offers the opportunity for dynamic control of pulse timing and shaping. This differs from the apparently similar induction motor which also has windings that are energised in a rotating phased sequence, in that the magnetization of the rotor is static (a salient pole that is made 'North` remains so as the motor rotates) while an induction motor has slip, and rotates at slightly less than synchronous speed. This absence of slip makes it possible to know the rotor position exactly, and the motor can be stepped arbitrarily slowly.

Simple Switching

If the poles A0 and A1 are energised then the rotor will align itself with these poles. Once this has occurred it is possible for the stator poles to be de-energised before the stator poles of B0 and B1 are energized. The rotor is now positioned at the stator poles b. This sequence continues through c before arriving back at the start. This sequence can also be reversed to achieve motion in the opposite direction. This sequence can be found to be unstable. while in operation, under high load, or high acceleration or deceleration, a step can be missed, and the rotor jumps to wrong angle, perhaps going back one instead of forward three.

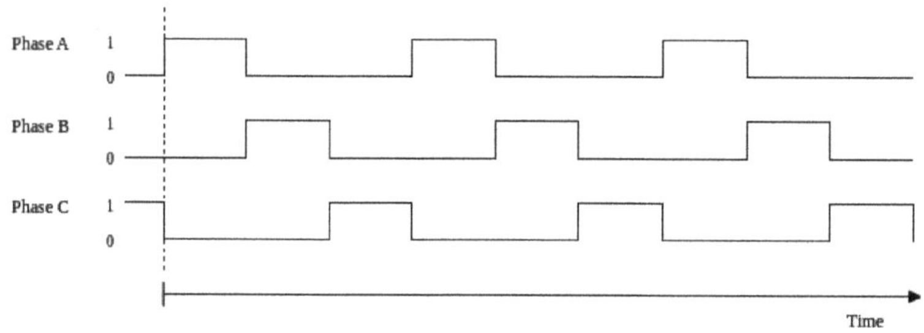

Improved Sequence

A much more stable system can be found by using the following "quadrature" sequence. First, stator poles A0 and A1 are energized. Then stator poles of B0 and B1 are energized which pulls the rotor so that it is aligned in between the stator poles of A and B. Following this the stator poles of A are de-energized and the rotor continues on to be aligned with the stator poles of B, this sequence continues through BC, C and CA before a full rotation has occurred. This sequence can also be reversed to achieve motion in the opposite direction. As at any time two coils are energised, and there are more steps between positions with identical magnetisation, so the onset of missed steps occurs at higher speeds or loads.

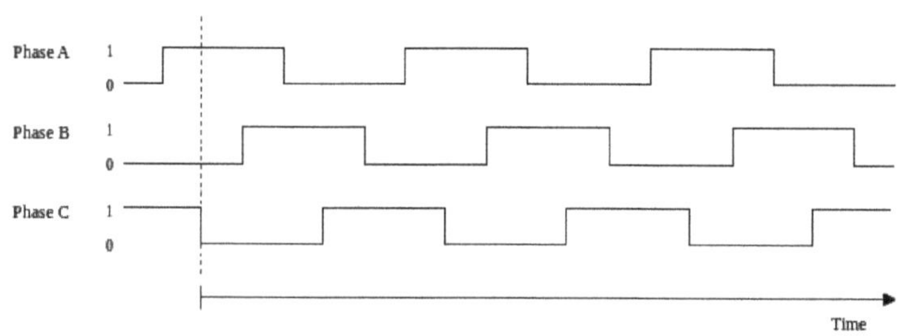

In addition to more stable operation, this approach leads to a duty cycle of each phase of 1/2, rather than 1/3 as in the simpler sequence.

Control

The control system is responsible for giving the required sequential pulses to the power circuitry in order to activate the phases as required. While it is possible to do this using electro-mechanical means such as commutators or simple analog or digital timing circuits, more control is possible with more advanced methods.

Many controllers in use incorporate programmable logic controllers (PLCs) rather than electromechanical components in their implementation. A microcontroller is also ideal for this kind of application since it enables a very precise control of the phase activation timings. It also gives the possibility of implementing a soft start function in software form, in order to reduce the amount of hardware required.

Power Circuitry

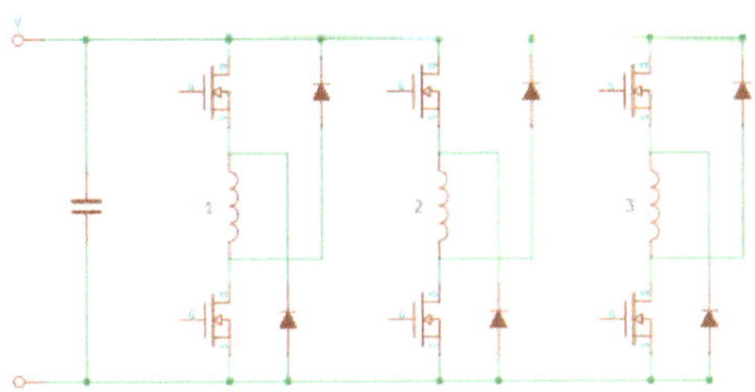

Asymmetric bridge converter

The most common approach to the powering of a switched reluctance motor is to use an asymmetric bridge converter.

There are 3 phases in an asymmetric bridge converter corresponding to the phases of the switched reluctance motor. If both of the power switches on either side of the phase are turned on, then that corresponding phase shall be actuated. Once the current has risen above the set value, the switch shall turn off. The energy now stored within the motor winding shall now maintain the current in the same direction until that energy is depleted.

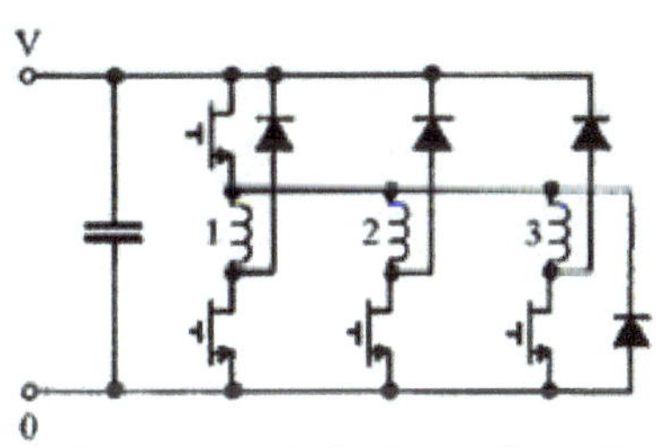

(n+1) switch and diode configuration

N+1 switch and diode

This basic circuitry may be altered so that fewer components are required although the circuit shall perform the same action. This efficient circuit is known as the (n+1) switch and diode configuration.

A capacitor, in either configuration, is used to suppress electrical and acoustic noise by limiting fluctuations in the supply voltage.

Universal AC-DC Motor

Modern low-cost universal motor, from a vacuum cleaner. Field windings are dark copper-colored, toward the back, on both sides. The rotor's laminated core is gray metallic, with dark slots for winding the coils. The commutator (partly hidden) has become dark from use; it is toward the front. The large brown molded-plastic piece in the foreground supports the brush guides and brushes (both sides), as well as the front motor bearing.

A commutated electrically excited series or parallel wound motor is referred to as a universal motor because it can be designed to operate on AC or DC power. A universal motor can operate well on AC because the current in both the field and the armature coils (and hence the resultant magnetic fields) will alternate (reverse polarity) in synchronism, and hence the resulting mechanical force will occur in a constant direction of rotation.

Operating at normal power line frequencies, universal motors are often found in a range less than 1000 watts. Universal motors also formed the basis of the traditional railway traction motor in electric railways. In this application, the use of AC to power a motor originally designed to run on DC would lead to efficiency losses due to eddy current heating of their magnetic components, particularly the motor field pole-pieces that, for DC, would have used solid (un-laminated) iron and they are now rarely used.

An advantage of the universal motor is that AC supplies may be used on motors which have some characteristics more common in DC motors, specifically high starting torque and very compact design if high running speeds are used. The negative aspect is the maintenance and short life problems caused by the commutator. Such motors are used in devices such as food mixers and power tools which are used only intermittently, and often have high starting-torque demands. Multiple taps on the field coil provide (imprecise) stepped speed control. Household blenders that advertise many speeds frequently combine a field coil with several taps and a diode that can be inserted in series with the motor (causing the motor to run on half-wave rectified AC). Universal motors also lend themselves to electronic speed control and, as such, are an ideal choice for devices like domestic washing machines. The motor can be used to agitate the drum (both forwards and in reverse) by switching the field winding with respect to the armature.

Whereas SCIMs cannot turn a shaft faster than allowed by the power line frequency, universal motors can run at much higher speeds. This makes them useful for appliances such as blenders, vacuum cleaners, and hair dryers where high speed and light weight are desirable. They are also commonly used in portable power tools, such as drills, sanders, circular and jig saws, where the motor's characteristics work well. Many vacuum cleaner and weed trimmer motors exceed 10,000 rpm, while many similar miniature grinders exceed 30,000 rpm.

UNIVERSAL MOTOR

The universal motor is so named because it is a type of electric motor that can operate on AC or DC power. It is a commutated series-wound motor where the stator's field coils are connected in series with the rotor windings through a commutator. It is often referred to as an AC series motor. The universal motor is very similar to a DC series motor in construction, but is modified slightly to allow the motor to operate properly on AC power. This type of electric motor can operate well on AC because the current in both the field coils and the armature (and the resultant magnetic fields) will alternate (reverse polarity) synchronously with the supply. Hence the resulting mechanical force will occur in a consistent direction of rotation, independent of the direction of applied voltage, but determined by the commutator and polarity of the field coils.

Universal motors have high starting torque, can run at high speed, and are lightweight and compact. They are commonly used in portable power tools and equipment, as well as many household appliances. They're also relatively easy to control, electromechanically using tapped coils, or electronically. However, the commutator has brushes that wear, so they are much less often used for equipment that is in continuous use. In addition, partly because of the commutator, universal motors are typically very noisy, both acoustically and electromagnetically.

Properties

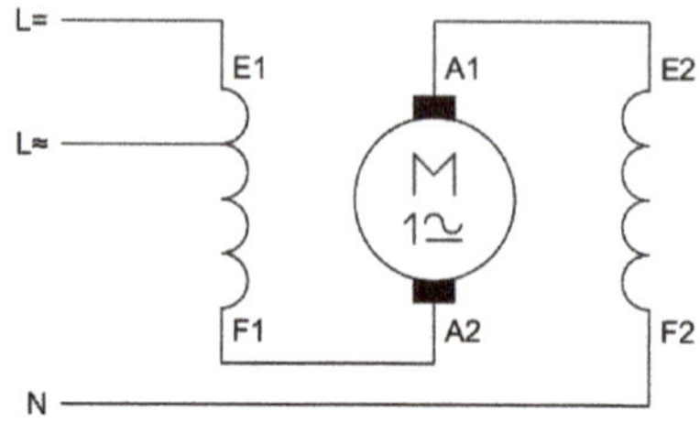

Universal motors' field coils are series wound with the rotor coils and commutator

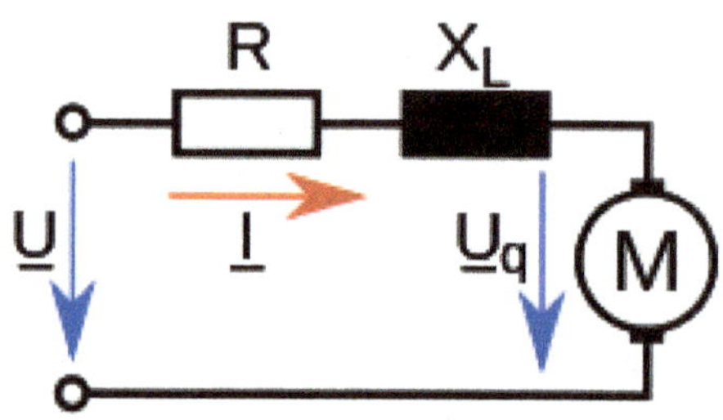

Equivalent circuit

Not all series wound motors operate well on AC current. If an ordinary series wound DC motor were connected to an AC supply, it would run very poorly. The universal motor is modified in several ways to allow for proper AC supply operation. There is a compensating winding typically added, along with laminated pole pieces, as opposed to the solid pole pieces found in DC motors. A universal motor's armature typically has far more coils and plates than a DC motor, and hence fewer windings per coil. This reduces the inductance.

Efficiency

Even when used with AC power these types of motors are able to run at a rotation frequency well above that of the mains supply, and because most electric motor properties improve with speed, this means they can be lightweight and powerful. However, universal motors are usually relatively inefficient- around 30% for smaller motors and up to 70-75% for larger ones.

Torque-Speed Characteristics

Series wound electric motors respond to increased load by slowing down; the current increases and the torque rises in proportion to the square of the current since the same current flows in both the armature and the field windings. If the motor is stalled, the current is limited only by the total resistance of the windings and the torque can be very high, and there is a danger of the windings becoming overheated. The counter-EMF aids the armature resistance to limit the current through the armature. When power is first applied to a motor, the armature does not rotate. At that instant, the counter-EMF is zero and the only factor limiting the armature current is the armature resistance. Usually the armature resistance of a motor is low; therefore the current through the armature would be very large when the power is applied. Therefore the need can arise for an additional resistance in series with the armature to limit the current until the motor rotation can build up the counter-EMF. As the motor rotation builds up, the resistance is gradually cut out. The speed-torque characteristic is an almost perfectly straight line between the stall torque and the no-load speed. This suits large inertial loads as the speed will drop until the motor slowly starts to rotate and these motors have a very high stalling torque.

As the speed increases, the inductance of the rotor means that the ideal commutating point changes. Small motors typically have fixed commutation. While some larger universal motors have rotatable commutation, this is rare. Instead larger universal motors often have *compensation windings* in series with the motor, or sometimes inductively coupled, and placed at ninety electrical degrees to the main field axis. These reduce the reactance of the armature, and improve the commutation.

One useful property of having the field windings in series with the armature winding is that as the speed increases the counter EMF naturally reduces the voltage across, and current through the field windings, giving field weakening at high speeds. This means that the motor has no theoretical maximum speed for any particular applied voltage. Universal motors can be and are generally run at high speeds, 4000-16000 rpm, and can go over 20,000 rpm. By way of contrast, AC synchronous and squirrel cage induction motors cannot turn a shaft faster than allowed by the power line frequency. In countries with 60 Hz(cycle/Sec) AC supply, this speed is limited to 3600 RPM.

Motor damage may occur from over-speeding (running at a rotational speed in excess of design limits) if the unit is operated with no significant mechanical load. On larger motors, sudden loss of load is to be avoided, and the possibility of such an occurrence is incorporated into the motor`s protection and control schemes. In some smaller applications, a fan blade attached to the shaft often acts as an artificial load to limit the motor speed to a safe level, as well as a means to circulate cooling airflow over the armature and field windings. If there were no mechanical limits placed on a universal motor it could theoretically speed out of control in the same way any series-wound DC motor can.

An advantage of the universal motor is that AC supplies may be used on motors which have some characteristics more common in DC motors, specifically high starting torque and very compact design if high running speeds are used.

Disadvantages

A negative aspect is the maintenance and short life problems caused by the commutator, as well as electromagnetic interference (EMI) issues due to any sparking. Because of the relatively high maintenance commutator brushes, universal motors are best-suited for devices such as food mixers and power tools which are used only intermittently, and often have high starting-torque demands.

Another negative aspect is that these motors may only be used where mostly-clean air is present at all times. Due to the dramatically increased risk of overheating, totally-enclosed fan cooled universal motors would be impractical, though some have been made. Such a motor would need a large fan to circulate enough air, decreasing efficiency since the motor must use more energy to cool itself. The impracticality comes from the resulting size, weight, and thermal management issues which open motors have none of.

Speed control

Continuous speed control of a universal motor running on AC is easily obtained by use of a thyristor circuit, while multiple taps on the field coil provide (imprecise) stepped speed control. Household blenders that advertise many speeds frequently combine a field coil with several taps and a diode that can be inserted in series with the motor (causing the motor to run on half-wave rectified AC).

Variations

Shunt Winding

Universal motors are series wound. Shunt winding was used experimentally, in the late 19th century, but was impractical owing to problems with commutation. Various schemes of embedded resistance, inductance and antiphase cross-coupling were attempted to reduce this. Universal motors, including shunt wound, were favoured as AC motors at this time as they were self-starting. When self-starting induction motors and automatic starters became available, these replaced the larger universal motors (above 1 hp) and the shunt wound.

Repulsion-Start

In the past, repulsion-start wound-rotor motors provided high starting torque, but with added complexity. Their rotors were similar to those of universal motors, but their brushes were connected only to each other. Transformer action induced current into the rotor. Brush position relative to field poles meant that starting torque was developed by rotor repulsion from the field poles. A centrifugal mechanism, when close to running speed, connected all commutator bars together to create the equivalent of a squirrel-cage rotor. As well, when close to operating speed, better motors lifted the brushes out of contact.

Applications

Domestic Appliances

Operating at normal power line frequencies, universal motors are often found in a range less than 1000 watts. Their high speed makes them useful for appliances such as blenders, vacuum cleaners, and hair dryers where high speed and light weight are desirable. They are also commonly used in portable power tools, such as drills, sanders, circular and jig saws, where the motor's characteristics work well. Many vacuum cleaner and weed trimmer motors exceed 10,000 RPM, while many Dremel and similar miniature grinders exceed 30,000 RPM.

Universal motors also lend themselves to electronic speed control and, as such, were an ideal choice for domestic washing machines. The motor can be used to agitate the drum (both forwards and in reverse) by switching the field winding with respect to the armature. The motor can also be run up to the high speeds required for the spin cycle. Nowadays, variable-frequency drive motors are more commonly used instead.

Rail Traction

Universal motors also formed the basis of the traditional railway traction motor in electric railways. In this application, the use of AC to power a motor originally designed to run on DC would lead to efficiency losses due to eddy current heating of their magnetic components, particularly the motor field pole-pieces that, for DC, would have used solid (non-laminated) iron. Although the heating effects are reduced by using laminated pole-pieces, as used for the cores of transformers and by the use of laminations of high permeability electrical steel, one solution available at the start of the 20th century was for the motors to be operated from very low frequency AC supplies, with 25 and 16 $^{2}/_{3}$ Hz (the latter subsequently shifted to 16.7 Hz, which is not subharmonic to 50 Hz mains) operation being common.

AC MOTOR

An **AC motor** is an electric motor driven by an alternating current (AC). The AC motor commonly consists of two basic parts, an outside stationary stator having coils supplied with alternating current to produce a rotating magnetic field, and an inside rotor attached to the output shaft producing a second rotating magnetic field. The rotor magnetic field may be produced by permanent magnets, reluctance saliency, or DC or AC electrical windings.

Less commonly, linear AC motors operate on similar principles as rotating motors but have their stationary and moving parts arranged in a straight line configuration, producing linear motion instead of rotation.

An industrial type of AC motor with electrical terminal box at the top and output rotating shaft on the left. Such motors are widely used for pumps, blowers, conveyors and other industrial machinery.

Operating Principles

When an AC motor is in steady-state rotation (motion), the magnetic fields of the rotor and stator rotate (move) with little or no slippage (near synchrony). The magnetic forces (repulsive and attractive) between the rotor and stator poles create average torque, capable of driving a load at rated speed. The speed of the stator rotating magnetic field (ω_s) and the speed of the rotor rotating magnetic field (ω_r), relative to the speed of the mechanical shaft (ω_m), must maintain synchronism for average torque production by satisfying the synchronous speed relation (i.e., $\pm\omega_s \pm \omega_r = \omega_m$) Otherwise, asynchronously rotating magnetic fields would produce pulsating or non-average torque.

The two main types of AC motors are classified as induction and synchronous. The induction motor (or asynchronous motor) always relies on a small difference in speed between the stator rotating magnetic field and the rotor shaft speed called slip to induce rotor current in the rotor AC winding. As a result, the induction motor cannot produce torque near synchronous speed where induction (or slip) is irrelevant or ceases to exist. In contrast, the synchronous motor does not rely on slip-induction for operation and uses either permanent magnets, salient poles (having projecting magnetic poles), or an independently excited rotor winding. The synchronous motor produces its rated torque at exactly synchronous speed. The brushless wound-rotor doubly fed synchronous motor system has an independently excited rotor winding that does not rely on the principles of slip-induction of current. The brushless wound-rotor doubly fed motor is a synchronous motor that can function exactly at the supply frequency or sub to super multiple of the supply frequency.

Other types of motors include eddy current motors, and also AC/DC mechanically commutated machines in which speed is dependent on voltage and winding connection.

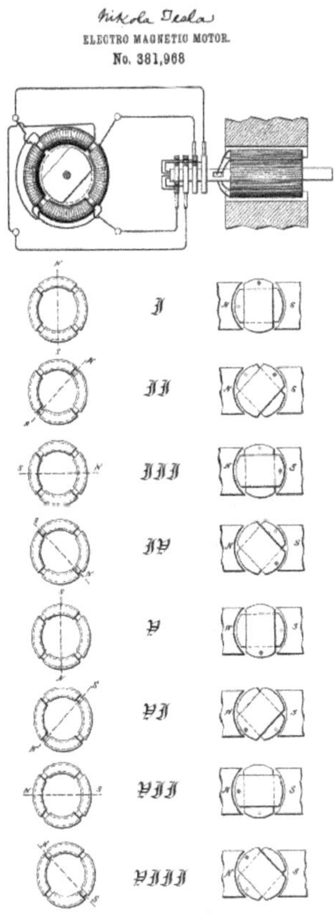

Drawing from U.S. Patent 381968, illustrating principle of Tesla's alternating current motor.

Alternating current technology was rooted in Michael Faraday's and Joseph Henry's 1830-31 discovery that a changing magnetic field can induce an electric current in a circuit. Faraday is usually given credit for this discovery since he published his findings first.

In 1832, French instrument maker Hippolyte Pixii generated a crude form of alternating current when he designed and built the first alternator. It consisted of a revolving horseshoe magnet passing over two wound wire coils.

Because of AC`s advantages in long distance high voltage transmission, there were many inventors in the United States and Europe during the late 19th century trying to develop workable AC motors. The first person to conceive of a rotating magnetic field was Walter Baily, who gave a workable demonstration of his battery-operated polyphase motor aided by a commutator on June 28, 1879 to the Physical Society of London. Describing an apparatus nearly identical to Baily's, French electrical engineer Marcel Deprez published a paper in 1880 that identified the rotating magnetic field principle and that of a two-phase AC system of currents to produce it. Never practically demonstrated, the design was flawed, as one of the two currents was "furnished by the machine itself." In 1886,

English engineer Elihu Thomson built an AC motor by expanding upon the induction-repulsion principle and his wattmeter. In 1887, American inventor Charles Schenk Bradley was the first to patent a two-phase AC power transmission with four wires.

"Commutatorless" alternating current induction motors seem to have been independently invented by Galileo Ferraris and Nikola Tesla. Ferraris demonstrated a working model of his single-phase induction motor in 1885, and Tesla built his working two-phase induction motor in 1887 and demonstrated it at the American Institute of Electrical Engineers in 1888 (although Tesla claimed that he conceived the rotating magnetic field in 1882). In 1888, Ferraris published his research to the Royal Academy of Sciences in Turin, where he detailed the foundations of motor operation; Tesla, in the same year, was granted a United States patent for his own motor. Working from Ferraris`s experiments, Mikhail Dolivo-Dobrovolsky introduced the first three-phase induction motor in 1890, a much more capable design that became the prototype used in Europe and the U.S. He also invented the first three-phase generator and transformer and combined them into the first complete AC three-phase system in 1891. The three-phase motor design was also worked on by the Swiss engineer Charles Eugene Lancelot Brown, and other three-phase AC systems were developed by German technician Friedrich August Haselwander and Swedish engineer Jonas Wenström.

Induction Motor

Cage and Wound Rotor Induction Motor

An induction motor is an asynchronous AC motor where power is transferred to the rotor by electromagnetic induction, much like transformer action. An induction motor resembles a rotating transformer, because the stator (stationary part) is essentially the primary side of the transformer and the rotor (rotating part) is the secondary side. Polyphase induction motors are widely used in industry.

Induction motors may be further divided into Squirrel Cage Induction Motors and Wound Rotor Induction Motors. SCIMs have a heavy winding made up of solid bars, usually aluminum or copper, joined by rings at the ends of the rotor. When one considers only the bars and rings as a whole, they are much like an animal's rotating exercise cage, hence the name.

Currents induced into this winding provide the rotor magnetic field. The shape of the rotor bars determines the speed-torque characteristics. At low speeds, the current induced in the squirrel cage is nearly at line frequency and tends to be in the outer parts of the rotor cage. As the motor accelerates, the slip frequency becomes lower, and more current is in the interior of the winding. By shaping the bars to change the resistance of the winding portions in the interior and outer parts of the cage, effectively a variable resistance is inserted in the rotor circuit. However, the majority of such motors have uniform bars.

In a WRIM, the rotor winding is made of many turns of insulated wire and is connected to slip rings on the motor shaft. An external resistor or other control devices can be connected in the rotor circuit. Resistors allow control of the motor speed, although significant power is dissipated in the external resistance. A converter can be fed from the rotor circuit and return the slip-frequency power that would otherwise be wasted back into the power system through an inverter or separate motor-generator.

The WRIM is used primarily to start a high inertia load or a load that requires a very high starting torque across the full speed range. By correctly selecting the resistors used in the secondary resistance or slip ring starter, the motor is able to produce maximum torque at a relatively low supply current from zero speed to full speed. This type of motor also offers controllable speed.

Motor speed can be changed because the torque curve of the motor is effectively modified by the amount of resistance connected to the rotor circuit. Increasing the value of resistance will move the speed of maximum torque down. If the resistance connected to the rotor is increased beyond the point where the maximum torque occurs at zero speed, the torque will be further reduced.

When used with a load that has a torque curve that increases with speed, the motor will operate at the speed where the torque developed by the motor is equal to the load torque. Reducing the load will cause the motor to speed up, and increasing the load will cause the motor to slow down until the load and motor torque are equal. Operated in this manner, the slip losses are dissipated in the secondary resistors and can be very significant. The speed regulation and net efficiency is also very poor.

Large 4,500 HP AC Induction Motor.

Torque Motor

A torque motor is a specialized form of electric motor which can operate indefinitely while stalled, that is, with the rotor blocked from turning, without incurring damage. In this mode of operation, the motor will apply a steady torque to the load (hence the name).

A common application of a torque motor would be the supply- and take-up reel motors in a tape drive. In this application, driven from a low voltage, the characteristics of these motors allow a relatively constant light tension to be applied to the tape whether or not the capstan is feeding tape past the tape heads. Driven from a higher voltage, (and so delivering a higher torque), the torque motors can also achieve fast-forward and rewind operation without requiring any additional mechanics such as gears or clutches. In the computer gaming world, torque motors are used in force feedback steering wheels.

Another common application is the control of the throttle of an internal combustion engine in conjunction with an electronic governor. In this usage, the motor works against a return spring to move the throttle in accordance with the output of the governor. The latter monitors engine speed by counting electrical pulses from the ignition system or from a magnetic pickup and, depending on the speed, makes small adjustments to the amount of current applied to the motor. If the engine starts to slow down relative to the desired speed, the current will be increased, the motor will develop more torque, pulling against the return spring and opening the throttle. Should the engine run too fast, the governor will reduce the current being applied to the motor, causing the return spring to pull back and close the throttle.

Synchronous Motor

A synchronous electric motor is an AC motor distinguished by a rotor spinning with coils passing magnets at the same rate as the AC and resulting magnetic field which drives it. Another way of saying this is that it has zero slip under usual operating conditions. Contrast this with an induction motor, which must slip to produce torque. One type of synchronous motor is like an induction motor except the rotor is excited by a DC field. Slip rings and brushes are used to conduct current to the rotor. The rotor poles connect to each other and move at the same speed hence the name synchronous motor. Another type, for low load torque, has flats ground onto a conventional squirrel-cage rotor to create discrete poles. Yet another, such as made by Hammond for its pre-World War II clocks, and in the older Hammond organs, has no rotor windings and discrete poles. It is not self-starting. The clock requires manual starting by a small knob on the back, while the older Hammond organs had an auxiliary starting motor connected by a spring-loaded manually operated switch.

Finally, hysteresis synchronous motors typically are (essentially) two-phase motors with a phase-shifting capacitor for one phase. They start like induction motors, but when slip rate decreases sufficiently, the rotor (a smooth cylinder) becomes temporarily magnetized. Its distributed poles make it act like a PMSM. The rotor material, like that of a common nail, will stay magnetized, but can also be demagnetized with little difficulty. Once running, the rotor poles stay in place; they do not drift.

Low-power synchronous timing motors (such as those for traditional electric clocks) may have multi-pole PM external cup rotors, and use shading coils to provide starting torque. *Telechron* clock motors have shaded poles for starting torque, and a two-spoke ring rotor that performs like a discrete two-pole rotor.

Doubly-Fed Electric Machine

Doubly fed electric motors have two independent multiphase winding sets, which contribute active (i.e., working) power to the energy conversion process, with at least one of the winding sets electronically controlled for variable speed operation. Two independent multiphase winding sets (i.e., dual armature) are the maximum provided in a single package without topology duplication. Doubly-fed electric motors are machines with an effective constant torque speed range that is twice synchronous speed for a given frequency of excitation. This is twice the constant torque speed range as singly-fed electric machines, which have only one active winding set.

A doubly-fed motor allows for a smaller electronic converter but the cost of the rotor winding and slip rings may offset the saving in the power electronics components. Difficulties with controlling speed near synchronous speed limit applications.

SPECIAL MAGNETIC MOTORS

Rotary

Ironless or Coreless Rotor Motor

A miniature coreless motor

Nothing in the principle of any of the motors described above requires that the iron (steel) portions of the rotor actually rotate. If the soft magnetic material of the rotor is made in the form of a cylinder, then (except for the effect of hysteresis) torque is exerted only on the windings of the electromagnets. Taking advantage of this fact is the *coreless or ironless DC motor*, a specialized form of a PM DC motor. Optimized for rapid acceleration, these motors have a rotor that is constructed without any iron core. The rotor can take the form of a winding-filled cylinder, or a self-supporting structure comprising only the magnet wire and the bonding material. The rotor can fit inside the stator magnets; a magnetically soft stationary cylinder inside the rotor provides a return path for the stator magnetic flux. A second arrangement has the rotor winding basket surrounding the stator magnets. In that design, the rotor fits inside a magnetically soft cylinder that can serve as the housing for the motor, and likewise provides a return path for the flux.

Because the rotor is much lighter in weight (mass) than a conventional rotor formed from copper windings on steel laminations, the rotor can accelerate much more rapidly, often achieving a mechanical time constant under one ms. This is especially true if the windings use aluminum rather than the heavier copper. But because there is no metal mass in the rotor to act as a heat sink, even small coreless motors must often be cooled by forced air. Overheating might be an issue for coreless DC motor designs. Modern software, such as Motor-CAD, can help to increase the thermal efficiency of motors while still in the design stage.

Among these types are the disc-rotor types, described in more detail in the next section.

Vibrator motors for cellular phones are sometimes tiny cylindrical PM field types, but there are also disc-shaped types which have a thin multipolar disc field magnet, and an intentionally unbalanced molded-plastic rotor structure with two bonded coreless coils. Metal brushes and a flat commutator switch power to the rotor coils.

Related limited-travel actuators have no core and a bonded coil placed between the poles of high-flux thin PMs. These are the fast head positioners for rigid-disk ("hard disk") drives. Although the contemporary design differs considerably from that of loudspeakers, it is still loosely (and incorrectly) referred to as a "voice coil" structure, because some earlier rigid-disk-drive heads moved in straight lines, and had a drive structure much like that of a loudspeaker.

Pancake or Axial Rotor Motor

A rather unusual motor design, the printed armature or pancake motor has the windings shaped as a disc running between arrays of high-flux magnets. The magnets are arranged in a circle facing the rotor with space in between to form an axial air gap. This design is commonly known as the pancake motor because of its extremely flat profile, although the technology has had many brand names since its inception, such as ServoDisc.

The printed armature (originally formed on a printed circuit board) in a printed armature motor is made from punched copper sheets that are laminated together using advanced composites to form a thin rigid disc. The printed armature has a unique construction in the brushed motor world in that it does not have a separate ring commutator. The brushes run directly on the armature surface making the whole design very compact.

An alternative manufacturing method is to use wound copper wire laid flat with a central conventional commutator, in a flower and petal shape. The windings are typically stabilized by being impregnated with electrical epoxy potting systems. These are filled epoxies that have moderate mixed viscosity and a long gel time. They are highlighted by low shrinkage and low exotherm, and are typically UL 1446 recognized as a potting compound insulated with 180 °C, Class H rating.

The unique advantage of ironless DC motors is that there is no cogging (torque variations caused by changing attraction between the iron and the magnets). Parasitic eddy currents cannot form in the rotor as it is totally ironless, although iron rotors are laminated. This can greatly improve efficiency, but variable-speed controllers must use a higher switching rate (>40 kHz) or DC because of the decreased electromagnetic induction.

These motors were originally invented to drive the capstan(s) of magnetic tape drives in the burgeoning computer industry, where minimal time to reach operating speed and minimal stopping distance were critical. Pancake motors are still widely used in high-performance servo-controlled systems, robotic systems, industrial automation and medical devices. Due to the variety of constructions now available, the technology is used in applications from high temperature military to low cost pump and basic servos.

Servo Motor

A servomotor is a motor, very often sold as a complete module, which is used within a position-control or speed-control feedback control system mainly control valves, such as motor-operated control valves. Servomotors are used in applications such as machine tools, pen plotters, and other process systems. Motors intended for use in a servomechanism must have well-documented characteristics for speed, torque, and power. The speed vs. torque curve is quite important and is high ratio for a servo motor. Dynamic response characteristics such as winding inductance and rotor inertia are also important; these factors limit the overall performance of the servomechanism loop. Large, powerful, but slow-responding servo loops may use conventional AC or DC motors and drive systems with position or speed feedback on the motor. As dynamic response requirements increase, more specialized motor designs such as coreless motors are used. AC motors' superior power density and acceleration characteristics compared to that of DC motors tends to favor PM synchronous, BLDC, induction, and SRM drive applications.

A servo system differs from some stepper motor applications in that the position feedback is continuous while the motor is running; a stepper system relies on the motor not to "miss steps" for short term accuracy, although a stepper system may include a "home" switch or other element to provide long-term stability of control. For instance, when a typical dot matrix computer printer starts up, its controller makes the print head stepper motor drive to its left-hand limit, where a position sensor defines home position and stops stepping. As long as power is on, a bidirectional counter in the printer's microprocessor keeps track of print-head position.

Stepper Motor

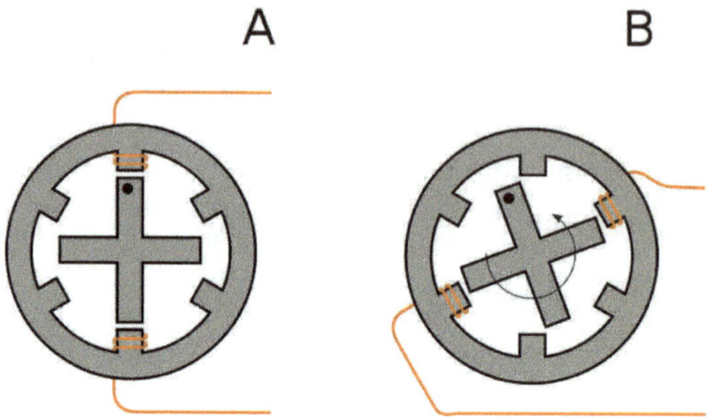

A stepper motor with a soft iron rotor, with active windings shown. In 'A' the active windings tend to hold the rotor in position. In 'B' a different set of windings are carrying a current, which generates torque and rotation.

Stepper motors are a type of motor frequently used when precise rotations are required. In a stepper motor an internal rotor containing PMs or a magnetically soft rotor with salient poles is controlled by a set of external magnets that are switched electronically. A stepper motor may also be thought of as a cross between a DC electric motor and a rotary solenoid. As each coil is energized in turn, the rotor aligns itself with the magnetic field produced by the energized field winding. Unlike a synchronous motor, in its application, the stepper motor may not rotate continuously; instead, it "steps"—starts and then quickly stops again—from one position to the next as field windings are energized and de-energized in sequence. Depending on the sequence, the rotor may turn forwards or backwards, and it may change direction, stop, speed up or slow down arbitrarily at any time.

Simple stepper motor drivers entirely energize or entirely de-energize the field windings, leading the rotor to "cog" to a limited number of positions; more sophisticated drivers can proportionally control the power to the field windings, allowing the rotors to position between the cog points and thereby rotate extremely smoothly. This mode of operation is often called microstepping. Computer controlled stepper motors are one of the most versatile forms of positioning systems, particularly when part of a digital servo-controlled system.

Stepper motors can be rotated to a specific angle in discrete steps with ease, and hence stepper motors are used for read/write head positioning in computer floppy diskette drives. They were used for the same purpose in pre-gigabyte era computer disk drives, where the precision and speed they offered was adequate for the correct positioning of the read/write head of a hard disk drive. As drive density increased, the precision and speed limitations of stepper motors made them obsolete for hard drives—the precision limitation made them unusable, and the speed limitation made them uncompetitive—thus newer hard disk drives use voice coil-based head actuator systems. (The term “voice coil” in this connection is historic; it refers to the structure in a typical (cone type) loudspeaker. This structure was used for a while to position the heads. Modern drives have a pivoted coil mount; the coil swings back and forth, something like a blade of a rotating fan. Nevertheless, like a voice coil, modern actuator coil conductors (the magnet wire) move perpendicular to the magnetic lines of force.)

Stepper motors were and still are often used in computer printers, optical scanners, and digital photocopiers to move the optical scanning element, the print head carriage (of dot matrix and inkjet printers), and the platen or feed rollers. Likewise, many computer plotters (which since the early 1990s have been replaced with large-format inkjet and laser printers) used rotary stepper motors for pen and platen movement; the typical alternatives here were either linear stepper motors or servomotors with closed-loop analog control systems.

So-called quartz analog wristwatches contain the smallest commonplace stepping motors; they have one coil, draw very little power, and have a PM rotor. The same kind of motor drives battery-powered quartz clocks. Some of these watches, such as chronographs, contain more than one stepping motor.

Closely related in design to three-phase AC synchronous motors, stepper motors and SRMs are classified as variable reluctance motor type.Stepper motors were and still are often used in computer printers, optical scanners, and computer numerical control (CNC) machines such as routers, plasma cutters and CNC lathes.

Linear Motor

A linear motor is essentially any electric motor that has been “unrolled” so that, instead of producing a torque (rotation), it produces a straight-line force along its length.

Linear motors are most commonly induction motors or stepper motors. Linear motors are commonly found in many roller-coasters where the rapid motion of the motorless railcar is controlled by the rail. They are also used in maglev trains, where the train “flies” over the ground. On a smaller scale, the 1978 era HP 7225A pen plotter used two linear stepper motors to move the pen along the X and Y axes.

Comparison by Major Categories

Type	Advantages	Disadvantages	Typical application	Typical drive, output
Self-commutated motors				
Brushed DC	Simple speed control Low initial cost	Maintenance (brushes) Medium lifespan Costly commutator and brushes	Steel mills Paper making machines Treadmill exercisers Automotive accessories	Rectifier, linear transistor(s) or DC chopper controller.

Brushless DC motor (BLDC) or (BLDM)	Long lifespan Low maintenance High efficiency	Higher initial cost Requires EC controller with closed-loop control	Rigid ("hard") disk drives CD/DVD players Electric vehicles RC Vehicles UAVs	Synchronous; single-phase or three-phase with PM rotor and trapezoidal stator winding; VFD typically VS PWM inverter type.
Switched reluctance motor (SRM)	Long lifespan Low maintenance High efficiency No permanent magnets Low cost Simple construction	Mechanical resonance possible High iron losses Not possible: * Open or vector control * Parallel operation Requires EC controller	Appliances Electric Vehicles Textile mills Aircraft applications	PWM and various other drive types, which tend to be used in very specialized / OEM applications.
Universal motor	High starting torque, compact, high speed.	Maintenance (brushes) Shorter lifespan Usually acoustically noisy Only small ratings are economical	Handheld power tools, blenders, vacuum cleaners, insulation blowers	Variable single phase AC, half-wave or full-wave phase-angle control with triac(s); closed-loop control optional.
AC asynchronous motors				
AC polyphase squirrel-cage or wound-rotor induction motor (SCIM) or (WRIM)	Self-starting Low cost Robust Reliable Ratings to 1+ MW Standardized types.	High starting current Lower efficiency due to need for magnetization.	Fixed-speed, traditionally, SCIM the world's workhorse especially in low performance applications of all types Variable-speed, traditionally, low-performance variable-torque pumps, fans, blowers and compressors. Variable-speed, increasingly, other high-performance constant-torque and constant-power or dynamic loads.	Fixed-speed, low performance applications of all types. Variable-speed, traditionally, WRIM drives or fixed-speed V/Hz-controlled VSDs. Variable-speed, increasingly, vector-controlled VSDs displacing DC, WRIM and single-phase AC induction motor drives.
AC SCIM split-phase capacitor-start	High power high starting torque	Speed slightly below synchronous Starting switch or relay required	Appliances Stationary Power Tools	Fixed or variable single-phase AC, variable speed being derived, typically, by full-wave phase-angle control with triac(s); closed-loop control optional.
AC SCIM split-phase capacitor-run	Moderate power High starting torque No starting switch Comparatively long life	Speed slightly below synchronous Slightly more costly	Industrial blowers Industrial machinery	
AC SCIM split-phase, auxiliary start winding	Moderate power Low starting torque	Speed slightly below synchronous Starting switch or relay required	Appliances Stationary power tools	
AC induction shaded-pole motor	Low cost Long life	Speed slightly below synchronous Low starting torque Small ratings low efficiency	Fans, appliances, record players	
AC synchronous motors				

Wound-rotor synchronous motor (WRSM)	Synchronous speed Inherently more efficient induction motor, low power factor	More costly	Industrial motors	Fixed or variable speed, three-phase; VFD typically six-step CS load-commutated inverter type or VS PWM inverter type.
Hysteresis motor	Accurate speed control Low noise No vibration High starting torque	Very low efficiency	Clocks, timers, sound producing or recording equipment, hard drive, capstan drive	Single-phase AC, two-phase capacitor-start, capacitor run motor
Synchronous reluctance motor (SyRM)	Equivalent to SCIM except more robust, more efficient, runs cooler, smaller footprint Competes with PM synchronous motor without demagnetization issues	Requires a controller Not widely available High cost	Appliances Electric vehicles Textile mills Aircraft applications	VFD can be standard DTC type or VS inverter PWM type.
Specialty motors				
Pancake or axial rotor motors	Compact design Simple speed control	Medium cost Medium lifespan	Office Equip Fans/Pumps, fast industrial and military servos	Drives can typically be brushed or brushless DC type.
Stepper motor	Precision positioning High holding torque	Some can be costly Require a controller	Positioning in printers and floppy disc drives; industrial machine tools	Not a VFD. Stepper position is determined by pulse counting.

Performance Parameters

Torque Capability of Motor Types

All the electromagnetic motors, and that includes the types mentioned here derive the torque from the vector product of the interacting fields. For calculating the torque it is necessary to know the fields in the air gap . Once these have been established by mathematical analysis using FEA or other tools the torque may be calculated as the integral of all the vectors of force multiplied by the radius of each vector. The current flowing in the winding is producing the fields and for a motor using a magnetic material the field is not linearly proportional to the current. This makes the calculation difficult but a computer can do the many calculations needed.

Once this is done a figure relating the current to the torque can be used as a useful parameter for motor selection. The maximum torque for a motor will depend on the maximum current although this will usually be only usable until thermal considerations take precedence.

When optimally designed within a given core saturation constraint and for a given active current (i.e., torque current), voltage, pole-pair number, excitation frequency (i.e., synchronous speed), and air-gap flux density, all categories of electric motors or generators will exhibit virtually the same maximum continuous shaft torque (i.e., operating torque) within a given air-gap area with winding slots and back-iron depth, which determines the physical size of electromagnetic core. Some applications require bursts of torque beyond the maximum operating torque, such as short bursts of torque to accelerate an electric vehicle from standstill. Always limited by magnetic core saturation or safe operating temperature rise and voltage, the capacity for torque bursts beyond the maximum operating torque differs significantly between categories of electric motors or generators.

Capacity for bursts of torque should not be confused with field weakening capability. Field weakening allows an electric machine to operate beyond the designed frequency of excitation. Field weakening is done when the maximum speed cannot be reached by increasing the applied voltage. This applies to only motors with current controlled fields and therefore cannot be achieved with PM motors.

Electric machines without a transformer circuit topology, such as that of WRSMs or PMSMs, cannot realize bursts of torque higher than the maximum designed torque without saturating the magnetic core and rendering any increase in current as useless. Furthermore, the PM assembly of PMSMs can be irreparably damaged, if bursts of torque exceeding the maximum operating torque rating are attempted.

Electric machines with a transformer circuit topology, such as induction machines, induction doubly-fed electric machines, and induction or synchronous wound-rotor doubly-fed (WRDF) machines, exhibit very high bursts of torque because the emf-induced active current on either side of the transformer oppose each other and thus contribute nothing to the transformer coupled magnetic core flux density, which would otherwise lead to core saturation.

Electric machines that rely on induction or asynchronous principles short-circuit one port of the transformer circuit and as a result, the reactive impedance of the transformer circuit becomes dominant as slip increases, which limits the magnitude of active (i.e., real) current. Still, bursts of torque that are two to three times higher than the maximum design torque are realizable.

The brushless wound-rotor synchronous doubly-fed (BWRSDF) machine is the only electric machine with a truly dual ported transformer circuit topology (i.e., both ports independently excited with no short-circuited port). The dual ported transformer circuit topology is known to be unstable and requires a multiphase slip-ring-brush assembly to propagate limited power to the rotor winding set. If a precision means were available to instantaneously control torque angle and slip for synchronous operation during motoring or generating while simultaneously providing brushless power to the rotor winding set, the active current of the BWRSDF machine would be independent of the reactive impedance of the transformer circuit and bursts of torque significantly higher than the maximum operating torque and far beyond the practical capability of any other type of electric machine would be realizable. Torque bursts greater than eight times operating torque have been calculated.

Continuous Torque Density

The continuous torque density of conventional electric machines is determined by the size of the air-gap area and the back-iron depth, which are determined by the power rating of the armature winding set, the speed of the machine, and the achievable air-gap flux density before core saturation. Despite the high coercivity of neodymium or samarium-cobalt PMs, continuous torque density is virtually the same amongst electric machines with optimally designed armature winding sets. Continuous torque density relates to method of cooling and permissible period of operation before destruction by overheating of windings or PM damage.

Continuous Power Density

The continuous power density is determined by the product of the continuous torque density and the constant torque speed range of the electric machine.

Standards

The following are major design, manufacturing, and testing standards covering electric motors:

- American Petroleum Institute: API 541 Form-Wound Squirrel Cage Induction Motors - 375 kW (500 Horsepower) and Larger
- American Petroleum Institute: API 546 Brushless Synchronous Machines - 500 kVA and Larger
- American Petroleum Institute: API 547 General-purpose Form-Wound Squirrel Cage Induction Motors - 250 Hp and Larger
- Institute of Electrical and Electronics Engineers: IEEE Std 112 Standard Test Procedure for Polyphase Induction Motors and Generators
- Institute of Electrical and Electronics Engineers: IEEE Std 115 Guide for Test Procedures for Synchronous Machines
- Institute of Electrical and Electronics Engineers: IEEE Std 841 Standard for Petroleum and Chemical Industry - Premium Efficiency Severe Duty Totally Enclosed Fan-Cooled (TEFC) Squirrel Cage Induction Motors - Up to and Including 370 kW (500 Hp)
- International Electrotechnical Commission: IEC 60034 Rotating Electrical Machines
- International Electrotechnical Commission: IEC 60072 Dimensions and output series for rotating electrical machines
- National Electrical Manufacturers Association: MG-1 Motors and Generators
- Underwriters Laboratories: UL 1004 - Standard for Electric Motors

Non-Magnetic Motors

An electrostatic motor is based on the attraction and repulsion of electric charge. Usually, electrostatic motors are the dual of conventional coil-based motors. They typically require a high-voltage power supply, although very small motors employ lower voltages. Conventional electric motors instead employ magnetic attraction and repulsion, and require high current at low voltages. In the 1750s, the first electrostatic motors were developed by Benjamin Franklin and Andrew Gordon. Today the electrostatic motor finds frequent use in micro-electro-mechanical systems (MEMS) where their drive voltages are below 100 volts, and where moving, charged plates are far easier to fabricate than coils and iron cores. Also, the molecular machinery which runs living cells is often based on linear and rotary electrostatic motors.

A piezoelectric motor or piezo motor is a type of electric motor based upon the change in shape of a piezoelectric material when an electric field is applied. Piezoelectric motors make use of the converse piezoelectric effect whereby the material produces acoustic or ultrasonic vibrations in order to produce a linear or rotary motion. In one mechanism, the elongation in a single plane is used to make a series stretches and position holds, similar to the way a caterpillar moves.

An electrically powered spacecraft propulsion system uses electric motor technology to propel spacecraft in outer space, most systems being based on electrically powering propellant to high speed, with some systems being based on electrodynamic tethers principles of propulsion to the magnetosphere.

REFERENCES

1. Faraday, Michael (1822). “On Some New Electro-Magnetical Motion, and on the Theory of Magnetism”. Quarterly Journal of Science, Literature and the Arts. Royal Institution of Great Britain. XII: 74–96 (§IX). Retrieved 12 February 2013.
2. Tom McInally, The Sixth Scottish University. The Scots Colleges Abroad: 1575 to 1799 (Brill, Leiden, 2012) p. 115
3. “The Development of the Electric Motor,”. Early Electric Motors. SparkMuseum. Retrieved 12 February 2013.
4. “The first dinamo?”. travelhungary.com. Retrieved 12 February 2013.
5. Guillemin, Amédée (1891). ‘Le Magnétisme et l’Électricitée’[Electricity and Magnetism]. trans., ed. & rev. from the French by Sylvanus P. Thompson. McMillan and Co.
6. Heller, Augustus (April 1896). “Anianus Jedlik”. Nature. Norman Lockyer. 53 (1379): 516. Bibcode:1896Natur..53..516H. doi:10.1038/053516a0.
7. Blundel, Stephen J. (2012). Magnetism A Very Short Introduction.Oxford University Press. p. 36. ISBN 978-0-19-960120-2.
8. Thein, M. “Elektrische Maschinen in Kraftfahrzeugen” [Electric Machines in Motor Vehicles] (PDF) (in German). Retrieved 13 February 2013.

9. "Elektrische Chronologie". Elektrisiermaschinen im 18. und 19. Jahrhundert – Ein kleines Lexikon ("Electrical machinery in the 18th and 19th centuries – a small thesaurus") (in German). University of Regensburg. March 31, 2004. Retrieved August 23, 2010.
10. "History of Batteries (inter alia)". Electropaedia. June 9, 2010. Retrieved August 23, 2010.
11. "Battery and Energy Technologies, Technology and Applications Timeline". Retrieved 13 February 2013.
12. Gee, William (2004). "Sturgeon, William (1783–1850)". Oxford Dictionary of National Biography. Oxford University Press. doi:10.1093/ref:odnb/26748.
13. Garrison, Ervan G. (1998). A History of Engineering and Technology: Artful Methods (2nd ed.). CRC Press. ISBN 0-8493-9810-X. Retrieved May 7, 2009.
14. Nye, David E. (1990). Electrifying America: Social Meanings of a New Technology. The MIT Press. ISBN 978-0-262-64030-5.
15. "Exhibition on the History of Hungarian Science". Retrieved 13 February 2013.
16. "Antonio Pacinotti".
17. "The Power Makers: Steam, Electricity, and the Men Who Invented Modern America".
18. "Zénobe Théophile Gramme". Invent Now, Inc. Hall of Fame profile. Retrieved 2012-09-19.
19. Ganot, Adolphe (1881). Elementary Treatise in Physics. Trans. and ed. from the French by E. Atkinson (14th ed.). William Wood and Co. pp. 907–908, sec. 899.
20. "Photo of a traditional form of the St. Louis motor".
21. "Buying an Energy-Efficient Electric Motor - Fact Sheet" (PDF). USDoE.
22. Babbage, C.; Herschel, J. F. W. (Jan 1825). "Account of the Repetition of M. Arago's Experiments on the Magnetism Manifested by Various Substances during the Act of Rotation". Philosophical Transactions of the Royal Society. 115 (0): 467–496. doi:10.1098/rstl.1825.0023. Retrieved 2 December 2012.
23. Thompson, Silvanus Phillips (1895). Polyphase Electric Currents and Alternate-Current Motors (1st ed.). London: E. & F.N. Spon. p. 261. Retrieved 2 December 2012.
24. Baily, Walter (June 28, 1879). "A Mode of Producing Arago's Rotation". Philosophical magazine: A journal of theoretical, experimental and applied physics. Taylor & Francis.
25. Vučković, Vladan (November 2006). "Interpretation of a Discovery" (PDF). The Serbian Journal of Electrical Engineers. 3 (2). Retrieved 10 February 2013.
26. Jonnes, Jill (2004). Empires of Light: Edison, Tesla, Westinghouse, and the Race to Electrify the World. Random House. p. 180.
27. Ferraris, G. (1888). "Atti della Reale Academia delle Science di Torino". Atti della R. Academia delle Science di Torino. XXIII: 360–375.

28. The Case Files: Nikola Tesla. “Two-Phase Induction Motor”. The Franklin Institute. Retrieved 2 December 2012.
29. “Galileo Ferraris Physicist, Pioneer of Alternating Current Systems (1847–1897) Inventor of the Induction Motor “Father of three-phase current” - Electrotechnical Congress, Frankfurt 1891”. Edison Tech Center. Retrieved 3 July 2012.
30. Smil, Vaclav (2005). Creating the Twentieth Century:Technical Innovations of 1867–1914 and Their Lasting Impact. Oxford University Press. p. 76. ISBN 978-0-19-988341-7.
31. Froehlich, Fritz E.; Kent, Allen (1 December 1998). The Froehlich/Kent Encyclopedia of Telecommunications: Volume 17 - Television Technology. CRC Press. pp. 37–. ISBN 978-0-8247-2915-8. Retrieved 10 October 2012.
32. Drury, Bill (2001). Control Techniques Drives and Controls Handbook. Institution of Electrical Engineers. p. xiv. ISBN 978-0-85296-793-5.
33. Langsdorf, Alexander Suss (1955). Theory of Alternating-Current Machinery (2nd ed.). Tata McGraw-Hill. p. 245. ISBN 0-07-099423-4.
34. The Encyclopedia Americana. 19. 1977. p. 518.
35. “Galileo Ferraris”. Encyclopædia Britannica.
36. “Biography of Galileo Ferraris”. Incredible People - Biographies of Famous People.
37. Neidhöfer, Gerhard. “Early Three-Phase Power (History)”. IEEE Power and Energy Magazine. 5 (5): 88–100. doi:10.1109/MPE.2007.904752.
38. Pansini, Anthony J. (1989). Basic of Electric Motors. Pennwell Publishing Company. p. 45. ISBN 0-13-060070-9.
39. Alger, P.L.; Arnold, R.E. (1976). “The History of Induction Motors in America”. Proceedings of the IEEE. 64 (9): 1380–1383. doi:10.1109/PROC.1976.10329.
40. Klooster, John W. (2009). Icons of Invention: The Makers of the Modern World from Gutenberg to Gates. ABC-CLIO, LLC. p. 305. ISBN 978-0-313-34746-7. Retrieved 10 September 2012.
41. Day, Lance; McNeil, Ian, eds. (1996). Biographical Dictionary of the History of Technology. London: Routledge. p. 1204. ISBN 0-203-02829-5. Retrieved 2 December 2012.
42. Froehlich, Fritz E. Editor-in-Chief; Allen Kent Co-Editor (1992). The Froehlich/Kent Encyclopedia of Telecommunications: Volume 17 - Television Technology to Wire Antennas (First ed.). New York: Marcel Dekker, Inc. p. 36. ISBN 0-8247-2902-1. Retrieved 2 December 2012.
43. The Electrical Engineer (21 Sep 1888). ... a new application of the AC in the production of rotary motion was made known almost simultaneously by two experimenters, Nikola Tesla and Galileo Ferraris, and the subject has attracted general attention from the fact that no commutator or connection of any kind with the armature was required ... Volume II. London: Charles & Co. p. 239.
44. Ferraris, Galileo (1885). “Electromagnetic Rotation with an Alternating Current”. Electrican. 36: 360–375.

45. Tesla, Nikola; AIEE Trans. (1888). "A New System for Alternating Current Motors and Transformers". AIEE. 5: 308–324. Retrieved 17 December 2012.
46. Harris, William (2008). "How Did Nikola Tesla Change the Way We Use Energy?". Howstuffworks.com. p. 3. Retrieved 10 September 2012.
47. Mattox, D. M. (2003). The Foundations of Vacuum Coating Technology. Random House. p. 39. ISBN 0-8155-1495-6.
48. Hughes, Thomas Parke. Networks of Power: Electrification in Western society, 1880–1930. Johns Hopkins University Press. p. 117. ISBN 0-8018-2873-2.
49. "Timeline of Nikola Tesla". Tesla Society of USA and Canada. Archived from the original on 8 May 2012. Retrieved 5 July 2012.
50. Hubbell, M.W. (2011). The Fundamentals of Nuclear Power Generation Questions & Answers. Authorhouse. p. 27. ISBN 978-1-4634-2441-1.
51. VDE Committee History of Electrical Engineering IEEE German Chapter (January 2012). "150th Birthday of Michael von Dolivo-Dobrowolsky Colloquium". 13. Retrieved 10 February 2013.
52. Dolivo-Dobrowolsky, M. (1891). ETZ. 12: 149, 161. Missing or empty |title= (help)
53. Mortensen, S. H.; Beckwith, S. (1949). "§7-1 'General Picture of a Synchronous Machine' in Sec. 7 - Alternating-Current Generators and Motors". In Knowlton, A.E. Standard Handbook for Electrical Engineers (8th ed.). McGraw-Hill. p. 646-647, figs. 7-1 & 7-2.
54. Hameyer, §5.1, p. 62
55. Lynn, §83, p. 812
56. "Variable Speed Pumping, A Guide to Successful Applications, Executive Summary" (PDF). USDOE - Europump - Hydraulic Institute. May 2004. p. 9, Fig. ES-7. Retrieved Feb 19, 2018. Check date values in: |access-date= (help)
57. Bose, Bimal K. (2006). "Chapters 6, 7 & 8". Power Electronics and Motor Drives : Advances and Trends. Academic Press. pp. see esp. 328, 397, 481. ISBN 978-0-12-088405-6.
58. Knight, Andy. Electric Machines. U. of Alberta. Retrieved 20 February 2013.
59. Hameyer, Kay (2001). "Electrical Machine I: Basics, Design, Function, Operation" (PDF). RWTH Aachen University Institute of Electrical Machines. Retrieved 11 January 2013.
60. Stölting, Hans-Dieter (2008). "§1.3.1.1 Motor Systematics in Chapter 1 - Introduction". In Stölting, H. D.; Kallenbach, E.; Amrhein, W. (eds.). Handbook of Fractional-Horsepower Drives (Online ed.). Springer. p. see esp. p. 5, Table 1.1. ISBN 978-3-540-73128-3.
61. Ionel, D.M. (2010). "High-Efficiency Variable-Speed Electric Motor Drive Technologies for Energy Savings in the US Residential Sector". 12th International Conference on Optimization of Electrical and Electronic Equipment (OPTIM). IEEE. pp. 1403–1414. doi:10.1109/OPTIM.2010.5510481.

62. Alger, Philip L.; et al. (1949). "§274-§287 'AC Commutator Motors' sub-section of Sec. 7 - Alternating-Current Generators and Motors". In Knowlton, A.E. Standard Handbook for Electrical Engineers (8th ed.). McGraw-Hill. pp. 755–763.
63. Krishnan, R. (2008). Permanent Magnet Synchronous and Brushless DC Motor Drives. CRC. p. xvii. ISBN 978-0-8247-5384-9.
64. Weiβmantel, H; Oesingmann, P.; Möckel, A. (2008). "§2.1 Motors with Commutator in Chapter 2 - Motors with Continuous Rotation". In Stölting, H. D.; Kallenbach, E.; Amrhein, W. (eds.). Handbook of Fractional-Horsepower Drives (Online ed.). Springer. pp. 13–160. ISBN 978-3-540-73128-3.
65. Liu, Chen-Ching; et al. (1997). "§66.1 'Generators' in Chapter 66 - 'Electric Machines' of Section 6 - 'Energy'". In Dorf, Richard C. (editor-in-chief). The Electrical Engineering Handbook (3rd ed.). CRC Press. p. 1456. ISBN 0-8493-8574-1.
66. Nozawa, Tetsuo (2009). "Tokai University Unveils 100W DC Motor with 96% Efficiency". Tech-On -- Nikkei Electronics.
67. Bush, Steve (2009). "Dyson vacuums 104,000 rpm brushless DC technology". Electronics Weekly Magazine. Archived from the original on 2012-04-11.
68. Lander, Cyril W. (1993). "§9–8 'Slip Ring Induction Motor Control' in Chapter 9 - A.C. Machine Control". Power electronics (3rd ed.). McGraw-Hill 480 pages. ISBN 0-07-707714-8.
69. Krishnan, R. (March 1987). "Selection Criteria for Servo Motor Drives". IEEE Transactions on Industry Applications. IA-23 (2): 270–275. doi:10.1109/TIA.1987.4504902. Retrieved 26 February 2013.
70. Patrick, Dale R.; Fardo, Stephen W. (1997). "Chapter 11". Rotating Electrical Machines and Power Systems (2nd ed.). Fairmont Press, Inc. ISBN 978-0-88173-239-9.
71. Bose, pp. 569–570, 891
72. Fenoglio, John A.; Chin, Bessie W. C.; Cobb, Terry R. (February 1979). "A High-Quality Digital X-Y Plotter Designed for Reliability, Flexibility and Low Cost" (PDF). Hewlett-Packard Journal. Retrieved 9 February 2012.
73. Stölting, p. 9
74. Bose, pp. 480–481
75. Vukosavic, Slobodan; Stefanovic, Victor R. (Nov–Dec 1991). "SRM Inverter Topologies: A Comparative Evaluation". IEEE Trans. on Industry Applications. 27 (6): 1034–1047. doi:10.1109/IAS.1990.152299.
76. Roters, Herbert C. (Jan 1947). "The hysteresis motor - Advances which permit economical fractional horsepower ratings". Trans. AIEE. 66 (1): 1419–1430. doi:10.1109/T-AIEE.1947.5059594.
77. Bakshi, U. A.; Bakshi, M. V. (2009). "§9.3 'Hysteresis Motors' in Chapter 9 - Special Machines". Electrical Machines - II (4th ed.). Technical Publications Pune. ISBN 81-8431-189-3.

78. Lendenmann, Heinz; et al. "Motoring Ahead" (PDF). Retrieved Apr 18, 2012.
79. Stölting, p. 10
80. Bose, p. 389
81. Kirtley, James L., Jr. (2005). "Class Notes 1: Electromagnetic Forces" (PDF). 6.6585 - Electric Machines. MIT Dept of Electrical Engineering. Retrieved 15 March 2013.
82. "DC Motor Calculations, part 1". National Instruments. 30 November 2011. Retrieved 7 December 2012.
83. Dwight, Herbert B.; Fink, D. G. (1949). "§27 to §35A Electromagnetic Induction of EMF in Sec. 2 - Electric and Magnetic Circuits". In Knowlton, A.E. Standard Handbook for Electrical Engineers (8th ed.). McGraw-Hill. pp. 36–41.
84. Laithwaite, E.R. (Feb 1975). "Linear electric machines - A personal view". Proceedings of the IEEE. 63 (2): 250–290. doi:10.1109/PROC.1975.9734.
85. Patterson, D. J.; Brice, C. W.; Dougal, R. A.; Kovuri, D. (1–4 June 2003). "The "Goodness" of Small Contemporary Permanent Magnet Electric Machines" (PDF). Electric Machines and Drives Conference, 2003. IEMDC'03. vol. 2. IEEE. pp. 1195–1200. doi:10.1109/IEMDC.2003.1210392.
86. Klatt, Frederick W. (Sep 2012). "Sensorless Real Time Control (RTC): Brushless Wound-Rotor [Synchronous] Doubly-Fed Machine". Sensorless Control for Electrical Drives -SLED 2012. IEEE. doi:10.1109/SLED.2012.6422811.
87. "Tethers Unlimited".
88. T.G. Wilson, P.H. Trickey, "D.C. Machine. With Solid State Commutation", AIEE paper I. CP62-1372, October 7, 1962
89. M. Gopal. Control systems: principles and design. 2nd ed. Tata McGraw-Hill, 2002. Page 165.
90. Curie temperature
91. "Brushless DC motor design", retrieved on April 26, 2015.
92. www.E BikeKit.com
93. ECMs and HVAC Systems
94. Ohio Electric Motors. Brushless DC Motors Used in Industrial Applications. Ohio Electric Motors. 2012. Archived November 8, 2012, at WebCite
95. Ohio Electric Motors. DC Motor Protection. Ohio Electric Motors. 2011. Archived November 28, 2011, at WebCite
96. Sabrie Soloman. Sensors handbook. 3rd ed. McGraw-Hill, 1999. Page 5-6.
97. Pter Campbel. Permanent Magnet Materials and their Application.Cambridge University Press, 1996. Page 172.
98. M. Gopal. Control systems: principles and design. 2nd ed. Tata McGraw-Hill, 2002. Page 159.
99. Shimon Y. Nof, Wilbert E. Wilhelm and Hans-Jürgen Warnecke. Industrial assembly. Springer, 1997. Page 174.

100. Peng Zhang. Industrial Control Technology: A Handbook for Engineers and Researchers. William Andrew, Inc., 2008. Page 91.
101. Ohio Electric Motors. Brushless DC Motors Used in Industrial Applications. Ohio Electric Motors. 2012. 1Archived November 8, 2012, at WebCite
102. Bernstein, Bobby (15 January 2015). "Top 4 Fastest RC Cars for Sale in the World". heavy.com. Retrieved 2 February 2015. As far as THE fastest RC car available for sale is concerned, it is the Traxxas XO-1 Supercar. The XO-1 hits 100mph, with proper LiPos batteries. The maker's product specs indicates the usage of a "Traxxas Big Block brushless motor"

CHAPTER 6

Transformer

Chapter Contents

- **Introduction**
- **Basic Principles**
- **Real Transformer**
- **Construction**

Electrical power transformer is a static device which transforms electrical energy from one circuit to another without any direct electrical connection and with the help of mutual induction between two windings. It transforms power from one circuit to another without changing its frequency but may be in different voltage level.

INTRODUCTION

A transformer is an electrical device that transfers electrical energy between two or more circuits through electromagnetic induction. Electromagnetic induction produces an electromotive force within a conductor which is exposed to time varying magnetic fields. Transformers are used to increase or decrease the alternating voltages in electric power applications.

A varying current in the transformer's primary winding creates a varying magnetic flux in the transformer core and a varying field impinging on the transformer's secondary winding. This varying magnetic field at the secondary winding induces a varying electromotive force (EMF) or voltage in the secondary winding due to electromagnetic induction. Making use of Faraday's Law (discovered in 1831) in conjunction with high magnetic permeability core properties, transformers can be designed to efficiently change AC voltages from one voltage level to another within power networks.

Since the invention of the first constant potential transformer in 1885, transformers have become essential for the transmission, distribution, and utilization of alternating current electrical energy. A wide range of transformer designs is encountered in electronic and electric power applications. Transformers range in size from RF transformers less than a cubic centimeter in volume to units interconnecting the power grid weighing hundreds of tons.

Pole-mounted distribution transformer with center-tapped secondary winding used to provide 'split-phase' power for residential and light commercial service, which in North America is typically rated 120/240 V.

BASIC PRINCIPLES

Ideal Transformer

For simplification or approximation purposes, it is very common to analyze the transformer as an ideal transformer model as presented in the two images. An ideal transformer is a theoretical, linear transformer that is lossless and perfectly coupled; that is, there are no energy losses and flux is completely confined within the magnetic core. Perfect coupling implies infinitely high core magnetic permeability and winding inductances and zero net magnetomotive force.

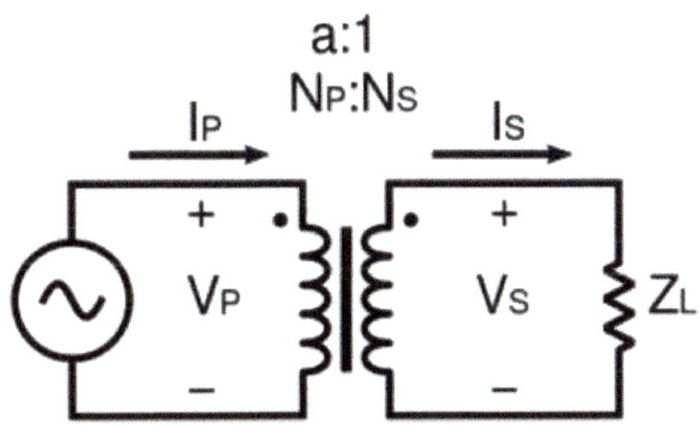

Ideal transformer connected with source V_P on primary and load impedance Z_L on secondary, where $0 < Z_L < \infty$.

A varying current in the transformer's primary winding creates a varying magnetic flux in the core and a varying magnetic field impinging on the secondary winding. This varying magnetic field at the secondary induces a varying electromotive force (EMF) or voltage in the secondary winding. The primary and secondary windings are wrapped around a core of infinitely high magnetic permeability so that all of the magnetic flux passes through both the primary and secondary windings. With a voltage source connected to the primary winding and load impedance connected to the secondary winding, the transformer currents flow in the indicated directions.

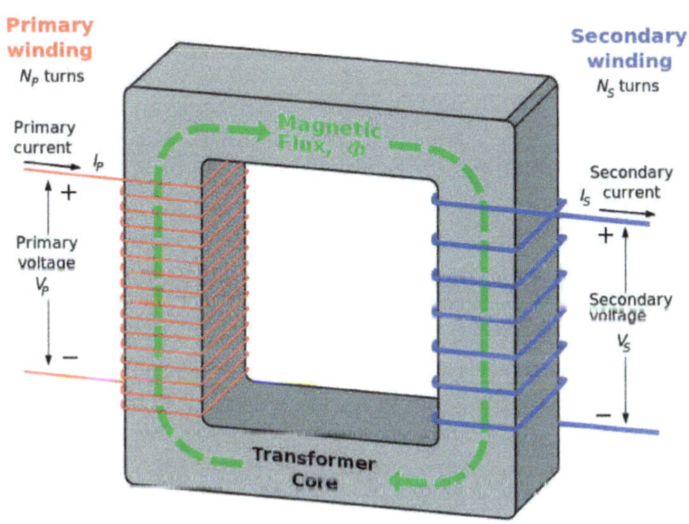

Ideal transformer and induction law[e]

According to Faraday's Law, since the same magnetic flux passes through both the primary and secondary windings in an ideal transformer, a voltage is induced in each winding. The primary EMF is sometimes termed counter EMF. This is in accordance with Lenz's law, which states that induction of EMF always opposes development of any such change in magnetic field.

The transformer winding voltage ratio is thus shown to be directly proportional to the winding turns ratio.

According to the law of conservation of energy, any load impedance connected to the ideal transformer's secondary winding results in conservation of apparent, real and reactive power consistent.

Instrument transformer, with polarity dot and X1 markings on LV side terminal

The ideal transformer identity reasonable approximation for the typical commercial transformer, with voltage ratio and winding turns ratio both being inversely proportional to the corresponding current ratio.

By Ohm's law and the ideal transformer identity:

- the secondary circuit load impedance can be expressed.
- the apparent load impedance referred to the primary circuit is derived. To be equal to the turns ratio squared times the secondary circuit load impedance.

Polarity

A dot convention is often used in transformer circuit diagrams, nameplates or terminal markings to define the relative polarity of transformer windings. Positively increasing instantaneous current entering the primary winding's dot end induces positive polarity voltage at the secondary winding's dot end.

REAL TRANSFORMER

Deviations from Ideal

The ideal transformer model neglects the following basic linear aspects in real transformers.

Core losses, collectively called magnetizing current losses, consist of

- Hysteresis losses due to nonlinear application of the voltage applied in the transformer core, and
- Eddy current losses due to joule heating in the core that are proportional to the square of the transformer's applied voltage.

Whereas windings in the ideal model have no resistances and infinite inductances, the windings in a real transformer have finite non-zero resistances and inductances associated with:

- Joule losses due to resistance in the primary and secondary windings
- Leakage flux that escapes from the core and passes through one winding only resulting in primary and secondary reactive impedance.

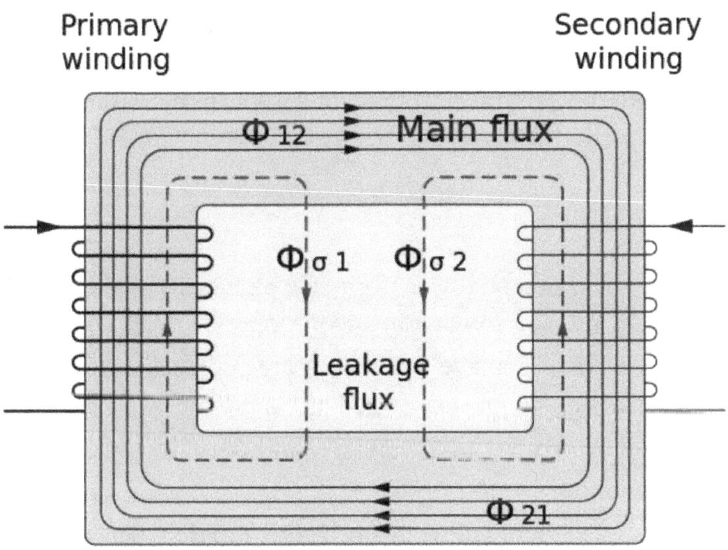

Leakage flux of a transformer

Leakage Flux

The ideal transformer model assumes that all flux generated by the primary winding links all the turns of every winding, including itself. In practice, some flux traverses paths that take it outside the windings. Such flux is termed leakage flux, and results in leakage inductance in series with the mutually coupled transformer windings. Leakage flux results in energy being alternately stored in and discharged from the magnetic fields with each cycle of the power supply. It is not directly a power loss, but results in inferior voltage regulation, causing the secondary voltage not to be directly proportional to the primary voltage, particularly under heavy load. Transformers are therefore normally designed to have very low leakage inductance.

In some applications increased leakage is desired, and long magnetic paths, air gaps, or magnetic bypass shunts may deliberately be introduced in a transformer design to limit the short-circuit current it will supply. Leaky transformers may be used to supply loads that exhibit negative resistance, such as electric arcs, mercury vapor lamps, and neon signs or for safely handling loads that become periodically short-circuited such as electric arc welders.

Air gaps are also used to keep a transformer from saturating, especially audio-frequency transformers in circuits that have a DC component flowing in the windings.

Knowledge of leakage inductance is also useful when transformers are operated in parallel. It can be shown that if the percent impedance[1] and associated winding leakage reactance-to-resistance (X/R) ratio of two transformers were hypothetically exactly the same, the transformers would share power in proportion to their respective volt-ampere ratings (e.g. 500 kVA unit in parallel with 1,000 kVA unit, the larger unit would carry twice the current). However, the impedance tolerances of commercial transformers are significant. Also, the Z impedance and X/R ratio of different capacity transformers tends to vary, corresponding 1,000 kVA and 500 kVA units' values being, to illustrate, respectively, $Z \approx 5.75\%$, $X/R \approx 3.75$ and $Z \approx 5\%$, $X/R \approx 4.75$.

Equivalent Circuit

Referring to the diagram, a practical transformer's physical behavior may be represented by an equivalent circuit model, which can incorporate an ideal transformer.

Winding joule losses and leakage reactances are represented by the following series loop impedances of the model:

- Primary winding: R_P, X_P
- Secondary winding: R_S, X_S.

In normal course of circuit equivalence transformation, R_S and X_S are in practice usually referred to the primary side by multiplying these impedances by the turns ratio squared, $(N_P/N_S)^2 = a^2$.

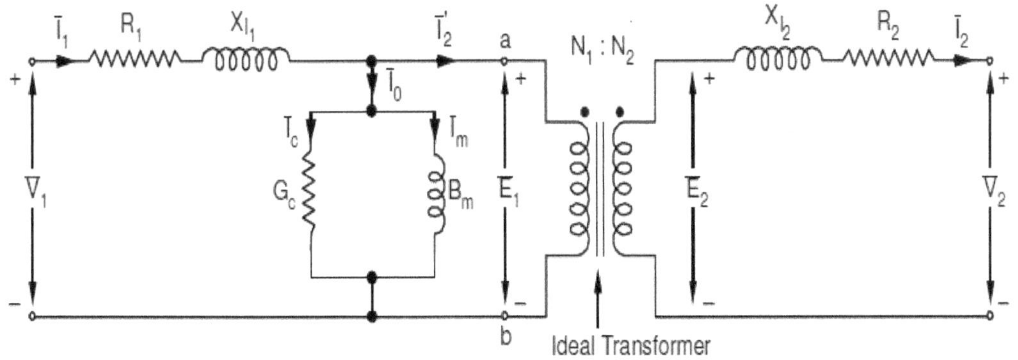

Real transformer equivalent circuit

Core loss and reactance is represented by the following shunt leg impedances of the model:

- Core or iron losses: R_C
- Magnetizing reactance: X_M.

R_C and X_M are collectively termed the magnetizing branch of the model.

Core losses are caused mostly by hysteresis and eddy current effects in the core and are proportional to the square of the core flux for operation at a given frequency. The finite permeability core requires a magnetizing current I_M to maintain mutual flux in the core. Magnetizing current is in phase with the flux, the relationship between the two being non-linear due to saturation effects. However, all impedances of the equivalent circuit shown are by definition linear and such non-linearity effects are not typically reflected in transformer equivalent circuits. With sinusoidal supply, core flux lags the induced EMF by 90°. With open-circuited secondary winding, magnetizing branch current I_0 equals transformer no-load current.

The resulting model, though sometimes termed 'exact' equivalent circuit based on linearity assumptions, retains a number of approximations.Analysis may be simplified by assuming that magnetizing branch impedance is relatively high and relocating the branch to the left of the primary impedances. This introduces error but allows combination of primary and referred secondary resistances and reactances by simple summation as two series impedances.

Transformer equivalent circuit impedance and transformer ratio parameters can be derived from the following tests: open-circuit test, short-circuit test, winding resistance test, and transformer ratio test.

Basic Transformer Parameters and Construction

Effect of Frequency

By Faraday's Law of induction transformer EMFs vary according to the derivative of flux with respect to time. The ideal transformer's core behaves linearly with time for any non-zero frequency. Flux in a real transformer's core behaves non-linearly in relation to magnetization current as the instantaneous flux increases beyond a finite linear range resulting in magnetic saturation associated with increasingly large magnetizing current, which eventually leads to transformer overheating.

The EMF of a transformer at a given flux density increases with frequency. By operating at higher frequencies, transformers can be physically more compact because a given core is able to transfer more power without reaching saturation and fewer turns are needed to achieve the same impedance. However, properties such as core loss and conductor skin effect also increase with frequency. Aircraft and military equipment employ 400 Hz power supplies which reduce core and winding weight. Conversely, frequencies used for some railway electrification systems were much lower (e.g. 16.7 Hz and 25 Hz) than normal utility frequencies (50–60 Hz) for historical reasons concerned mainly with the limitations of early electric traction motors. Consequently, the transformers used to step-down the high overhead line voltages (e.g. 15 kV) were much larger and heavier for the same power rating than those required for the higher frequencies.

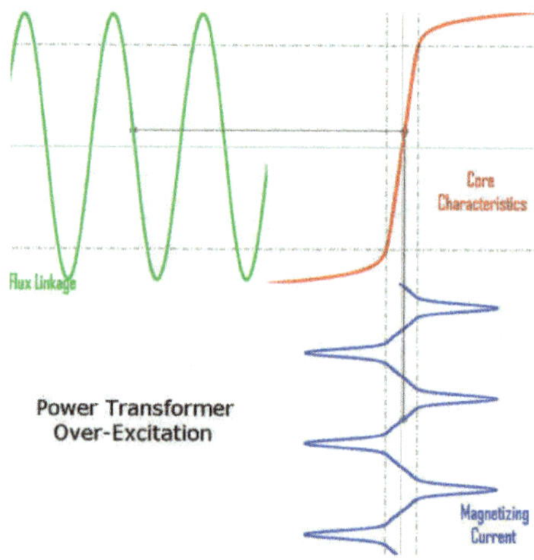

Power transformer over-excitation condition caused by decreased frequency; flux (green), iron core's magnetic characteristics (red) and magnetizing current (blue).

Operation of a transformer at its designed voltage but at a higher frequency than intended will lead to reduced magnetizing current. At a lower frequency, the magnetizing current will increase. Operation of a transformer at other than its design frequency may require assessment of voltages, losses, and cooling to establish if safe operation is practical. For example, transformers may need to be equipped with 'volts per hertz' over-excitation relays to protect the transformer from overvoltage at higher than rated frequency.

One example is in traction transformers used for electric multiple unit and high-speed train service operating across regions with different electrical standards. The converter equipment and traction transformers have to accommodate different input frequencies and voltage (ranging from as high as 50 Hz down to 16.7 Hz and rated up to 25 kV) while being suitable for multiple AC asynchronous motor and DC converters and motors with varying harmonics mitigation filtering requirements.

Large power transformers are vulnerable to insulation failure due to transient voltages with high-frequency components, such as caused in switching or by lightning.

At much higher frequencies the transformer core size required drops dramatically: a physically small and cheap transformer can handle power levels that would require a massive iron core at mains frequency. The development of switching power semiconductor devices and complex integrated circuits made switch-mode power supplies viable, to generate a high frequency from a much lower one (or DC), change the voltage level with a small transformer, and, if necessary, rectify the changed voltage

Energy Losses

Real transformer energy losses are dominated by winding resistance joule and core losses. Transformers' efficiency tends to improve with increasing transformer capacity. The efficiency of typical distribution transformers is between about 98 and 99 percent.

As transformer losses vary with load, it is often useful to express these losses in terms of no-load loss, full-load loss, half-load loss, and so on. Hysteresis and eddy current losses are constant at all load levels and dominate overwhelmingly without load, while variable winding joule losses dominating increasingly as load increases. The no-load loss can be significant, so that even an idle transformer constitutes a drain on the electrical supply. Designing energy efficient transformers for lower loss requires a larger core, good-quality silicon steel, or even amorphous steel for the core and thicker wire, increasing initial cost. The choice of construction represents a trade-off between initial cost and operating cost.

Transformer losses arise from:

Winding Joule Losses

Current flowing through a winding's conductor causes joule heating. As frequency increases, skin effect and proximity effect causes the winding's resistance and, hence, losses to increase.

Hysteresis Losses

Each time the magnetic field is reversed, a small amount of energy is lost due to hysteresis within the core. According to Steinmetz's formula, the heat energy due to hysteresis is given by

$$W_h \approx \eta \beta_{max}^{1.6}, \text{ and,}$$

hysteresis loss is thus given by

$$P_h \approx W_h f \approx \eta f \beta_{max}^{1.6}$$

where, f is the frequency, η is the hysteresis coefficient and β_{max} is the maximum flux density, the empirical exponent of which varies from about 1.4 to 1.8 but is often given as 1.6 for iron.

Eddy Current Losses

Ferromagnetic materials are also good conductors and a core made from such a material also constitutes a single short-circuited turn throughout its entire length. Eddy currents therefore circulate within the core in a plane normal to the flux, and are responsible for resistive heating of the core material. The eddy current loss is a complex function of the square of supply frequency and inverse square of the material thickness. Eddy current losses can be reduced by making the core of a stack of plates electrically insulated from each other, rather than a solid block; all transformers operating at low frequencies use laminated or similar cores.

Magnetostriction Related Transformer Hum

Magnetic flux in a ferromagnetic material, such as the core, causes it to physically expand and contract slightly with each cycle of the magnetic field, an effect known as magnetostriction, the frictional energy of which produces an audible noise known as mains hum or transformer hum. This transformer hum is especially objectionable in transformers supplied at power frequencies[o] and in high-frequency flyback transformers associated with television CRTs.

Stray Losses

Leakage inductance is by itself largely lossless, since energy supplied to its magnetic fields is returned to the supply with the next half-cycle. However, any leakage flux that intercepts nearby conductive materials such as the transformer's support structure will give rise to eddy currents and be converted to heat. There are also radiative losses due to the oscillating magnetic field but these are usually small.

Mechanical Vibration and Audible Noise Transmission

In addition to magnetostriction, the alternating magnetic field causes fluctuating forces between the primary and secondary windings. This energy incites vibration transmission in interconnected metalwork, thus amplifying audible transformer hum.

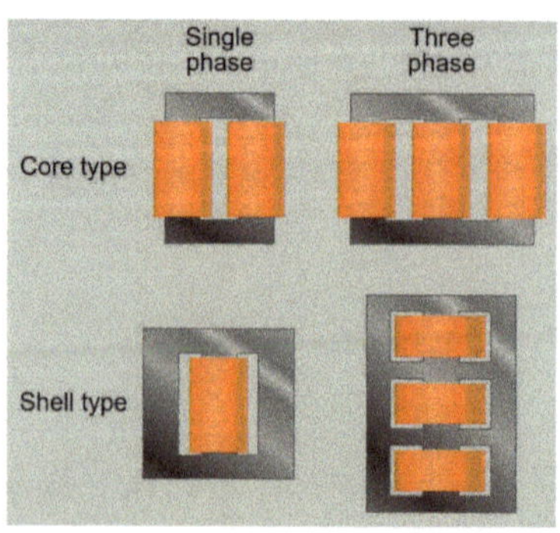

Core form = core type; shell form = shell type

Core form and Shell form Transformers

Closed-core transformers are constructed in 'core form' or 'shell form'. When windings surround the core, the transformer is core form; when windings are surrounded by the core, the transformer is shell form. Shell form design may be more prevalent than core form design for distribution transformer applications due to the relative ease in stacking the core around winding coils. Core form design tends to, as a general rule, be more economical, and therefore more prevalent, than shell form design for high voltage power transformer applications at the lower end of their voltage and power rating ranges (less than or equal to, nominally, 230 kV or 75 MVA). At higher voltage and power ratings, shell form transformers tend to be more prevalent. Shell form design tends to be preferred for extra-high voltage and higher MVA applications because, though more labor-intensive to manufacture, shell form transformers are characterized as having inherently better kVA-to-weight ratio, better short-circuit strength characteristics and higher immunity to transit damage.

CONSTRUCTION

Cores

Laminated Steel Cores

Laminated core transformer showing edge of laminations at top of photo

Power transformer inrush current caused by residual flux at switching instant; flux (green), iron core's magnetic characteristics (red) and magnetizing current (blue).

Transformers for use at power or audio frequencies typically have cores made of high permeability silicon steel. The steel has a permeability many times that of free space and the core thus serves to greatly reduce the magnetizing current and confine the flux to a path which closely couples the windings. Early transformer developers soon realized that cores constructed from solid iron resulted in prohibitive eddy current losses, and their designs mitigated this effect with cores consisting of bundles of insulated iron wires. Later designs constructed the core by stacking layers of thin steel laminations, a principle that has remained in use. Each lamination is insulated from its neighbors by a thin non-conducting layer of insulation. The universal transformer equation indicates a minimum cross-sectional area for the core to avoid saturation.

The effect of laminations is to confine eddy currents to highly elliptical paths that enclose little flux, and so reduce their magnitude. Thinner laminations reduce losses, but are more laborious and expensive to construct. Thin laminations are generally used on high-frequency transformers, with some of very thin steel laminations able to operate up to 10 kHz.

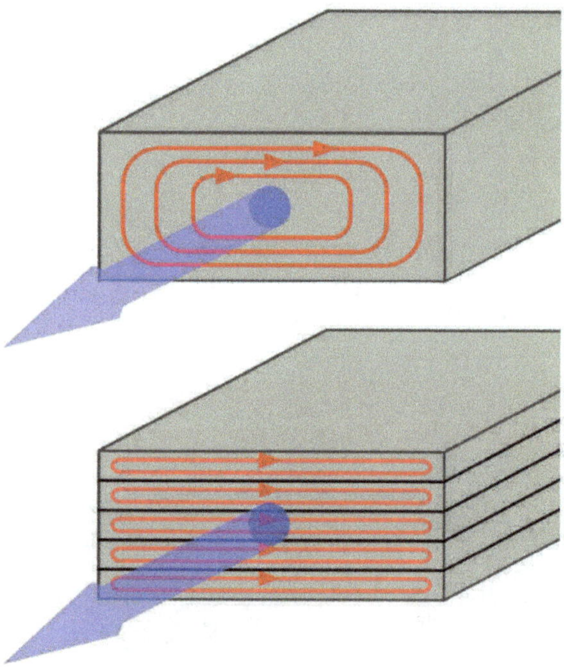

Laminating the core greatly reduces eddy-current losses

One common design of laminated core is made from interleaved stacks of E-shaped steel sheets capped with I-shaped pieces, leading to its name of 'E-I transformer'.Such a design tends to exhibit more losses, but is very economical to manufacture. The cut-core or C-core type is made by winding a steel strip around a rectangular form and then bonding the layers together. It is then cut in two, forming two C shapes, and the core assembled by binding the two C halves together with a steel strap. They have the advantage that the flux is always oriented parallel to the metal grains, reducing reluctance.

A steel core's remanence means that it retains a static magnetic field when power is removed. When power is then reapplied, the residual field will cause a high inrush current until the effect of the remaining magnetism is reduced, usually after a few cycles of the applied AC waveform. Overcurrent protection devices such as fuses must be selected to allow this harmless inrush to pass. On transformers connected to long, overhead power transmission lines, induced currents due to geomagnetic disturbances during solar storms can cause saturation of the core and operation of transformer protection devices.

Distribution transformers can achieve low no-load losses by using cores made with low-loss high-permeability silicon steel or amorphous (non-crystalline) metal alloy. The higher initial cost of the core material is offset over the life of the transformer by its lower losses at light load.

Solid Cores

Powdered iron cores are used in circuits such as switch-mode power supplies that operate above mains frequencies and up to a few tens of kilohertz. These materials combine high magnetic permeability with high bulk electrical resistivity. For frequencies extending beyond the VHF band, cores made from non-conductive magnetic ceramic materials called ferrites are common. Some radio-frequency transformers also have movable cores (sometimes called 'slugs') which allow adjustment of the coupling coefficient (and bandwidth) of tuned radio-frequency circuits.

Toroidal Cores

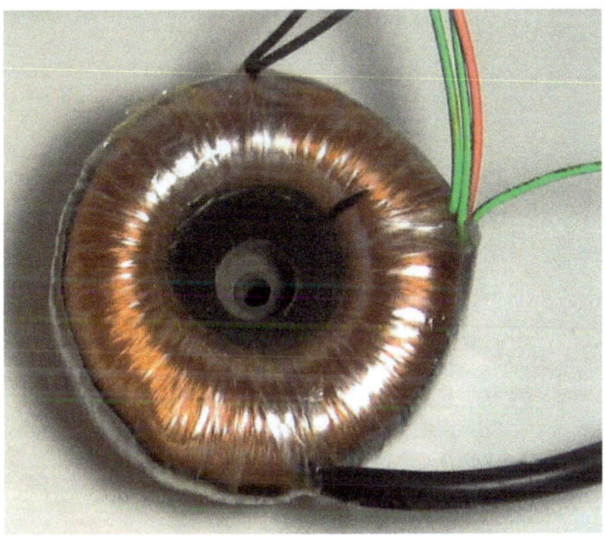

Small toroidal core transformer

Toroidal transformers are built around a ring-shaped core, which, depending on operating frequency, is made from a long strip of silicon steel or permalloy wound into a coil, powdered iron, or ferrite. A strip construction ensures that the grain boundaries are optimally aligned, improving the transformer's efficiency by reducing the core's reluctance. The closed ring shape eliminates air gaps inherent in the construction of an E-I core. The cross-section of the ring is usually square or rectangular, but more expensive cores with circular cross-sections are also available. The primary and secondary coils are often wound concentrically to cover the entire surface of the core. This minimizes the length of wire needed and provides screening to minimize the core's magnetic field from generating electromagnetic interference.

Toroidal transformers are more efficient than the cheaper laminated E-I types for a similar power level. Other advantages compared to E-I types, include smaller size (about half), lower weight (about half), less mechanical hum (making them superior in audio amplifiers), lower exterior magnetic field (about one tenth), low off-load losses (making them more efficient in standby circuits), single-bolt mounting, and greater choice of shapes. The main disadvantages are higher cost and limited power capacity (see Classification parameters below). Because of the lack of a residual gap in the magnetic path, toroidal transformers also tend to exhibit higher inrush current, compared to laminated E-I types.

Ferrite toroidal cores are used at higher frequencies, typically between a few tens of kilohertz to hundreds of megahertz, to reduce losses, physical size, and weight of inductive components. A drawback of toroidal transformer construction is the higher labor cost of winding. This is because it is necessary to pass the entire length of a coil winding through the core aperture each time a single turn is added to the coil. As a consequence, toroidal transformers rated more than a few kVA are uncommon. Relatively few toroids are offered with power ratings above 10 kVA, and practically none above 25 kVA. Small distribution transformers may achieve some of the benefits of a toroidal core by splitting it and forcing it open, then inserting a bobbin containing primary and secondary windings.

Air Cores

A physical core is not an absolute requisite and a functioning transformer can be produced simply by placing the windings near each other, an arrangement termed an 'air-core' transformer. The air which comprises the magnetic circuit is essentially lossless, and so an air-core transformer eliminates loss due to hysteresis in the core material. The leakage inductance is inevitably high, resulting in very poor regulation, and so such designs are unsuitable for use in power distribution. They have however very high bandwidth, and are frequently employed in radio-frequency applications, for which a satisfactory coupling coefficient is maintained by carefully overlapping the primary and secondary windings. They're also used for resonant transformers such as Tesla coils where they can achieve reasonably low loss in spite of the high leakage inductance.

Windings

The conducting material used for the windings depends upon the application, but in all cases the individual turns must be electrically insulated from each other to ensure that the current travels throughout every turn. For small power and signal transformers, in which currents are low and the potential difference between adjacent turns is small, the coils are often wound from enamelled magnet wire, such as Formvar wire. Larger power transformers operating at high voltages may be wound with copper rectangular strip conductors insulated by oil-impregnated paper and blocks of pressboard.

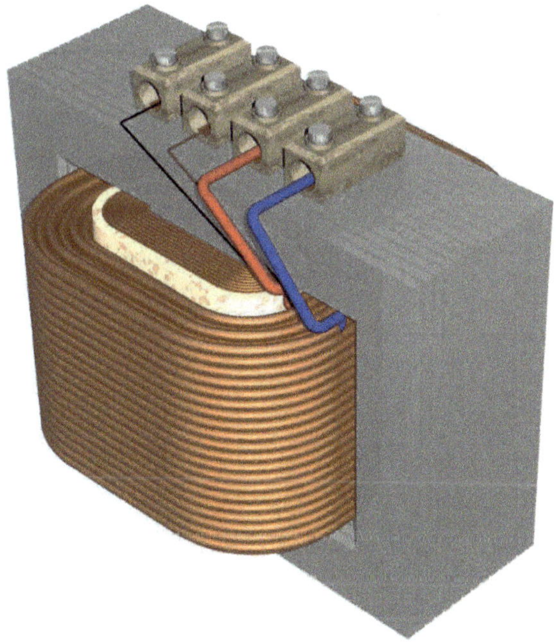

Windings are usually arranged concentrically to minimize flux leakage.

High-frequency transformers operating in the tens to hundreds of kilohertz often have windings made of braided Litz wire to minimize the skin-effect and proximity effect losses. Large power transformers use multiple-stranded conductors as well, since even at low power frequencies non-uniform distribution of current would otherwise exist in high-current windings. Each strand is individually insulated, and the strands are arranged so that at certain points in the winding, or throughout the whole winding, each portion occupies different relative positions in the complete conductor. The transposition equalizes the current flowing in each strand of the conductor, and reduces eddy current losses in the winding itself. The stranded conductor is also more flexible than a solid conductor of similar size, aiding manufacture.

The windings of signal transformers minimize leakage inductance and stray capacitance to improve high-frequency response. Coils are split into sections, and those sections interleaved between the sections of the other winding.

Power-frequency transformers may have taps at intermediate points on the winding, usually on the higher voltage winding side, for voltage adjustment. Taps may be manually reconnected, or a manual or automatic switch may be provided for changing taps. Automatic on-load tap changers are used in electric power transmission or distribution, on equipment such as arc furnace transformers, or for automatic voltage regulators for sensitive loads. Audio-frequency transformers, used for the distribution of audio to public address loudspeakers, have taps to allow adjustment of impedance to each speaker. A center-tapped transformer is often used in the output stage of an audio power amplifier in a push-pull circuit. Modulation transformers in AM transmitters are very similar.

Dry-type transformer winding insulation systems can be either of standard open-wound 'dip-and-bake' construction or of higher quality designs that include vacuum pressure impregnation (VPI), vacuum pressure encapsulation (VPE), and cast coil encapsulation processes. In the VPI process, a combination of heat, vacuum and pressure is used to thoroughly seal, bind, and eliminate entrained air voids in the winding polyester resin insulation coat layer, thus increasing resistance to corona. VPE windings are similar to VPI windings but provide more protection against environmental effects, such as from water, dirt or corrosive ambients, by multiple dips including typically in terms of final epoxy coat.

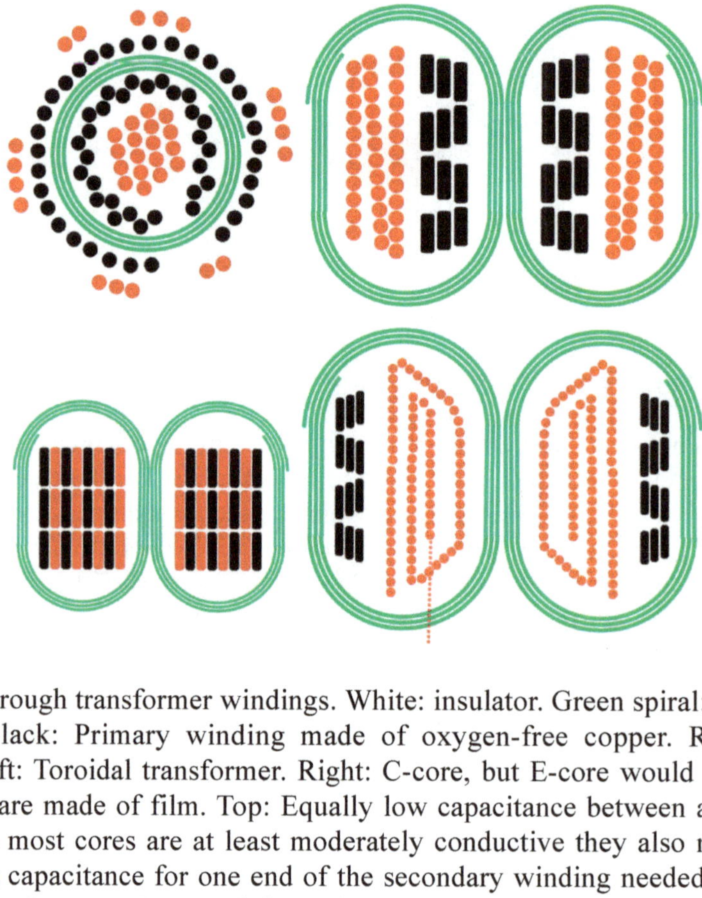

Cut view through transformer windings. White: insulator. Green spiral: Grain oriented silicon steel. Black: Primary winding made of oxygen-free copper. Red: Secondary winding. Top left: Toroidal transformer. Right: C-core, but E-core would be similar. The black windings are made of film. Top: Equally low capacitance between all ends of both windings. Since most cores are at least moderately conductive they also need insulation. Bottom: Lowest capacitance for one end of the secondary winding needed for low-power high-voltage transformers. Bottom left: Reduction of leakage inductance would lead to increase of capacitance.

Cooling

Cutaway view of liquid-immersed construction transformer. The conservator (reservoir) at top provides liquid-to-atmosphere isolation as coolant level and temperature changes. The walls and fins provide required heat dissipation balance.

To place the cooling problem in perspective, the accepted rule of thumb is that the life expectancy of insulation in all electrics, including all transformers, is halved for about every 7 °C to 10 °C increase in operating temperature, this life expectancy halving rule holding more narrowly when the increase is between about 7 °C to 8 °C in the case of transformer winding cellulose insulation.

Small dry-type and liquid-immersed transformers are often self-cooled by natural convection and radiation heat dissipation. As power ratings increase, transformers are often cooled by forced-air cooling, forced-oil cooling, water-cooling, or combinations of these. Large transformers are filled with transformer oil that both cools and insulates the windings.Transformer oil is a highly refined mineral oil that cools the windings and insulation by circulating within the transformer tank. The mineral oil and paper insulation system has been extensively studied and used for more than 100 years. It is estimated that 50% of power transformers will survive 50 years of use, that the average age of failure of power transformers is about 10 to 15 years, and that about 30% of power transformer failures are due to insulation and overloading failures. Prolonged operation at elevated temperature degrades insulating properties of winding insulation and dielectric coolant, which not only shortens transformer life but can ultimately lead to catastrophic transformer failure. With a great body of empirical study as a guide, transformer oil testing including dissolved gas analysis provides valuable maintenance information. This underlines the need to monitor, model, forecast and manage oil and winding conductor insulation temperature conditions under varying, possibly difficult, power loading conditions.

Building regulations in many jurisdictions require indoor liquid-filled transformers to either use dielectric fluids that are less flammable than oil, or be installed in fire-resistant rooms. Air-cooled dry transformers can be more economical where they eliminate the cost of a fire-resistant transformer room.

The tank of liquid filled transformers often has radiators through which the liquid coolant circulates by natural convection or fins. Some large transformers employ electric fans for forced-air cooling, pumps for forced-liquid cooling, or have heat exchangers for water-cooling. An oil-immersed transformer may be equipped with a Buchholz relay, which, depending on severity of gas accumulation due to internal arcing, is used to either alarm or de-energize the transformer. Oil-immersed transformer installations usually include fire protection measures such as walls, oil containment, and fire-suppression sprinkler systems.

Polychlorinated biphenyls have properties that once favored their use as a dielectric coolant, though concerns over their environmental persistence led to a widespread ban on their use. Today, non-toxic, stable silicone-based oils, or fluorinated hydrocarbons may be used where the expense of a fire-resistant liquid offsets additional building cost for a transformer vault. PCBs for new equipment were banned in 1981 and in 2000 for use in existing equipment in United KingdomLegislation enacted in Canada between 1977 and 1985 essentially bans PCB use in transformers manufactured in or imported into the country after 1980, the maximum allowable level of PCB contamination in existing mineral oil transformers being 50 ppm.

Some transformers, instead of being liquid-filled, have their windings enclosed in sealed, pressurized tanks and cooled by nitrogen or sulfur hexafluoride gas.

Experimental power transformers in the 500-to-1,000 kVA range have been built with liquid nitrogen or helium cooled superconducting windings, which eliminates winding losses without affecting core losses.

Insulation Drying

Construction of oil-filled transformers requires that the insulation covering the windings be thoroughly dried of residual moisture before the oil is introduced. Drying is carried out at the factory, and may also be required as a field service. Drying may be done by circulating hot air around the core, or by vapor-phase drying (VPD) where an evaporated solvent transfers heat by condensation on the coil and core.

For small transformers, resistance heating by injection of current into the windings is used. The heating can be controlled very well, and it is energy efficient. The method is called low-frequency heating (LFH) since the current used is at a much lower frequency than that of the power grid, which is normally 50 or 60 Hz. A lower frequency reduces the effect of inductance, so the voltage required can be reduced. The LFH drying method is also used for service of older transformers.

Bushings

Larger transformers are provided with high-voltage insulated bushings made of polymers or porcelain. A large bushing can be a complex structure since it must provide careful control of the electric field gradient without letting the transformer leak oil.

Classification Parameters

Transformers can be classified in many ways, such as the following:

- Power capacity: From a fraction of a volt-ampere (VA) to over a thousand MVA.
- Duty of a transformer: Continuous, short-time, intermittent, periodic, varying.
- Frequency range: Power-frequency, audio-frequency, or radio-frequency.
- Voltage class: From a few volts to hundreds of kilovolts.
- Cooling type: Dry and liquid-immersed – self-cooled, forced air-cooled; liquid-immersed – forced oil-cooled, water-cooled.
- Circuit application: Such as power supply, impedance matching, output voltage and current stabilizer or circuit isolation.
- Utilization: Pulse, power, distribution, rectifier, arc furnace, amplifier output, etc..
- Basic magnetic form: Core form, shell form.
- Constant-potential transformer descriptor: Step-up, step-down, isolation.
- General winding configuration: By EIC vector group – various possible two-winding combinations of the phase designations delta, wye or star, and zigzag or interconnected star; other – autotransformer, Scott-T, zigzag grounding transformer winding.
- Rectifier phase-shift winding configuration: 2-winding, 6-pulse; 3-winding, 12-pulse; . . . n-winding, [n-1] 6-pulse; polygon; etc..

Types

Various specific electrical application designs require a variety of transformer types. Although they all share the basic characteristic transformer principles, they are customize in construction or electrical properties for certain installation requirements or circuit conditions.

- Autotransformer: Transformer in which part of the winding is common to both primary and secondary circuits, leading to increased efficiency, smaller size, and a higher degree of voltage regulation.

- Capacitor voltage transformer: Transformer in which capacitor divider is used to reduce high voltage before application to the primary winding.
- Distribution transformer, power transformer: International standards make a distinction in terms of distribution transformers being used to distribute energy from transmission lines and networks for local consumption and power transformers being used to transfer electric energy between the generator and distribution primary circuits.
- Phase angle regulating transformer: A specialised transformer used to control the flow of real power on three-phase electricity transmission networks.
- Scott-T transformer: Transformer used for phase transformation from three-phase to two-phase and vice versa.
- Polyphase transformer: Any transformer with more than one phase.
- Grounding transformer: Transformer used for grounding three-phase circuits to create a neutral in a three wire system, using a wye-delta transformer, or more commonly, a zigzag grounding winding.
- Leakage transformer: Transformer that has loosely coupled windings.
- Resonant transformer: Transformer that uses resonance to generate a high secondary voltage.
- Audio transformer: Transformer used in audio equipment.
- Output transformer: Transformer used to match the output of a valve amplifier to its load.
- Instrument transformer: Potential or current transformer used to accurately and safely represent voltage, current or phase position of high voltage or high power circuits.
- Pulse transformer: Specialized small-signal transformer used to transmit digital signaling while providing electrical isolation, commonly used in Ethernet computer networks as 10BASE-T, 100BASE-T and 1000BASE-T.

An electrical substation in Melbourne, Australia showing three of five 220 kV – 66 kV transformers, each with a capacity of 150 MVA

Applications

Transformer at the Limestone Generating Station in Manitoba, Canada

Transformers are used to increase (or step-up) voltage before transmitting electrical energy over long distances through wires. Wires have resistance which loses energy through joule heating at a rate corresponding to square of the current. By transforming power to a higher voltage transformers enable economical transmission of power and distribution. Consequently, transformers have shaped the electricity supply industry, permitting generation to be located remotely from points of demand. All but a tiny fraction of the world's electrical power has passed through a series of transformers by the time it reaches the consumer.

Since the high voltages carried in the wires are significantly greater than what is needed in-home, transformers are also used extensively in electronic products to decrease (or step-down) the supply voltage to a level suitable for the low voltage circuits they contain. The transformer also electrically isolates the end user from contact with the supply voltage.

Signal and audio transformers are used to couple stages of amplifiers and to match devices such as microphones and record players to the input of amplifiers. Audio transformers allowed telephone circuits to carry on a two-way conversation over a single pair of wires. A balun transformer converts a signal that is referenced to ground to a signal that has balanced voltages to ground, such as between external cables and internal circuits. Transformers made to medical grade standards isolate the users from the direct current. These are found commonly used in conjunction with hospital beds, dentist chairs, and other medical lab equipment.

Discovery of Induction

Electromagnetic induction, the principle of the operation of the transformer, was discovered independently by Michael Faraday in 1831, Joseph Henry in 1832, and others. The relationship between EMF and magnetic flux is an equation now known as Faraday's law of induction:

$$|\mathcal{E}| = \left|\frac{d\Phi_B}{dt}\right|.$$

Where $|\mathcal{E}|$ is the magnitude of the EMF in Volts and Φ_B is the magnetic flux through the circuit in webers.

Faraday performed early experiments on induction between coils of wire, including winding a pair of coils around an iron ring, thus creating the first toroidal closed-core transformer. However he only applied individual pulses of current to his transformer, and never discovered the relation between the turns ratio and EMF in the windings.

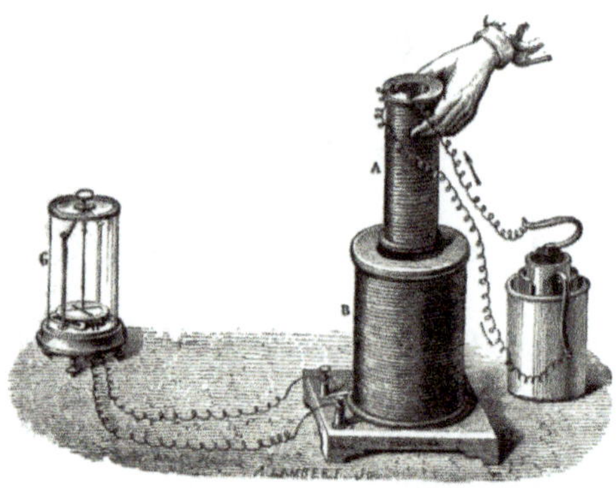

Faraday's experiment with induction between coils of wire

Induction Coils

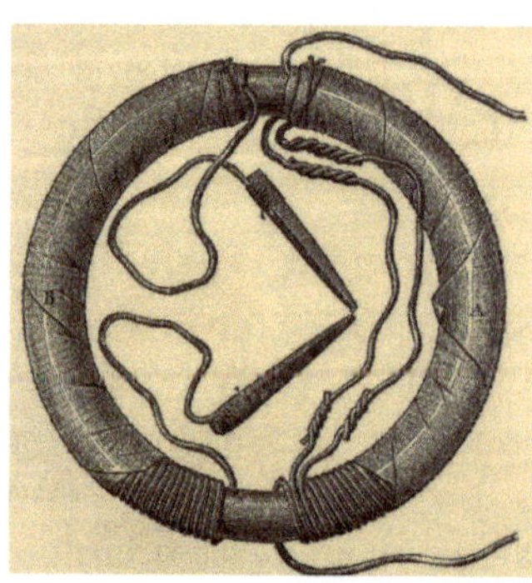

Faraday's ring transformer

The first type of transformer to see wide use was the induction coil, invented by Rev. Nicholas Callan of Maynooth College, Ireland in 1836. He was one of the first researchers to realize the more turns the secondary winding has in relation to the primary winding, the larger the induced secondary EMF will be. Induction coils evolved from scientists' and inventors' efforts to get higher voltages from batteries. Since batteries produce direct current (DC) rather than AC, induction coils relied upon vibrating electrical contacts that regularly interrupted the current in the primary to create the flux changes necessary for induction. Between the 1830s and the 1870s, efforts to build better induction coils, mostly by trial and error, slowly revealed the basic principles of transformers.

Induction coil, 1900, Bremerhaven, Germany

First Alternating Current Transformers

By the 1870s, efficient generators producing alternating current (AC) were available, and it was found AC could power an induction coil directly, without an interrupter.

In 1876, Russian engineer Pavel Yablochkov invented a lighting system based on a set of induction coils where the primary windings were connected to a source of AC. The secondary windings could be connected to several 'electric candles' (arc lamps) of his own design. The coils Yablochkov employed functioned essentially as transformers.

In 1878, the Ganz factory, Budapest, Hungary, began equipment for electric lighting and, by 1883, had installed over fifty systems in Austria-Hungary. Their AC systems used arc and incandescent lamps, generators, and other equipment.

Lucien Gaulard and John Dixon Gibbs first exhibited a device with an open iron core called a 'secondary generator' in London in 1882, then sold the idea to the Westinghouse company in the United States. They also exhibited the invention in Turin, Italy in 1884, where it was adopted for an electric lighting system. However, the efficiency of their open-core bipolar apparatus remained very low.

Early Series Circuit Transformer Distribution

Induction coils with open magnetic circuits are inefficient at transferring power to loads. Until about 1880, the paradigm for AC power transmission from a high voltage supply to a low voltage load was a series circuit. Open-core transformers with a ratio near 1:1 were connected with their primaries in series to allow use of a high voltage for transmission while presenting a low voltage to the lamps. The inherent flaw in this method was that turning off a single lamp (or other electric device) affected the voltage supplied to all others on the same circuit. Many adjustable transformer designs were introduced to compensate for this problematic characteristic of the series circuit, including those employing methods of adjusting the core or bypassing the magnetic flux around part of a coil. Efficient, practical transformer designs did not appear until the 1880s, but within a decade, the transformer would be instrumental in the War of Currents, and in seeing AC distribution systems triumph over their DC counterparts, a position in which they have remained dominant ever since.

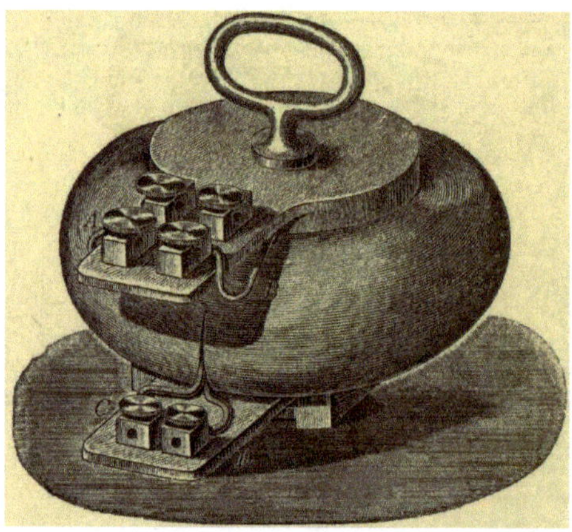

Shell form transformer. Sketch used by Uppenborn to describe ZBD engineers' 1885 patents and earliest articles.

Core form, front; shell form, back. Earliest specimens of ZBD-designed high-efficiency constant-potential transformers manufactured at the Ganz factory in 1885.

The ZBD team consisted of Károly Zipernowsky, Ottó Bláthy and Miksa Déri

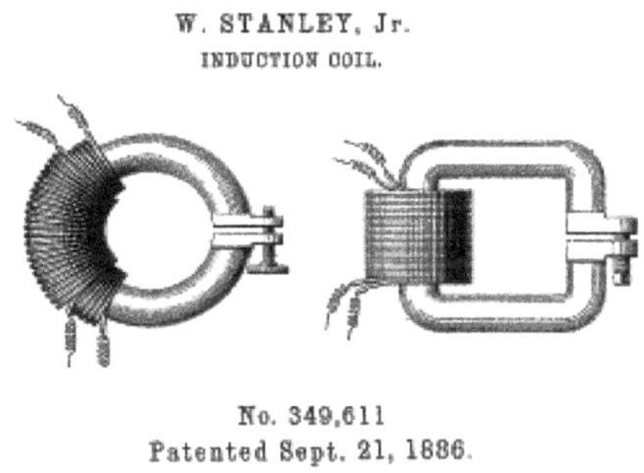

Stanley's 1886 design for adjustable gap open-core induction coils

Closed-Core Transformers and Parallel Power Distribution

In the autumn of 1884, Károly Zipernowsky, Ottó Bláthy and Miksa Déri (ZBD), three engineers associated with the Ganz factory, had determined that open-core devices were impracticable, as they were incapable of reliably regulating voltage. In their joint 1885 patent applications for novel transformers (later called ZBD transformers), they described two designs with closed magnetic circuits where copper windings were either a) wound around iron wire ring core or b) surrounded by iron wire core. The two designs were the first application of the two basic transformer constructions in common use to this day, which can as a class all be termed as either core form or shell form (or alternatively, core type or shell type), as in a) or b), respectively (see images). The Ganz factory had also in the autumn of 1884 made delivery of the world's first five high-efficiency AC transformers, the first of these units having been shipped on September 16, 1884. This first unit had been manufactured to the following specifications: 1,400 W, 40 Hz, 120:72 V, 11.6:19.4 A, ratio 1.67:1, one-phase, shell form.

In both designs, the magnetic flux linking the primary and secondary windings traveled almost entirely within the confines of the iron core, with no intentional path through air (see Toroidal cores below). The new transformers were 3.4 times more efficient than the open-core bipolar devices of Gaulard and Gibbs. The ZBD patents included two other major interrelated innovations: one concerning the use of parallel connected, instead of series connected, utilization loads, the other concerning the ability to have high turns ratio transformers such that the supply network voltage could be much higher (initially 1,400 to 2,000 V) than the voltage of utilization loads (100 V initially preferred). When employed in parallel connected electric distribution systems, closed-core transformers finally made it technically and economically feasible to provide electric power for lighting in homes, businesses and public spaces. Bláthy had suggested the use of closed cores, Zipernowsky had suggested the use of parallel shunt connections, and Déri had performed the experiments;

Transformers today are designed on the principles discovered by the three engineers. They also popularized the word 'transformer' to describe a device for altering the EMF of an electric current, although the term had already been in use by 1882. In 1886, the ZBD engineers designed, and the Ganz factory supplied electrical equipment for, the world's first power station that used AC generators to power a parallel connected common electrical network, the steam-powered Rome-Cerchi power plant.

Although George Westinghouse had bought Gaulard and Gibbs' patents in 1885, the Edison Electric Light Company held an option on the US rights for the ZBD transformers, requiring Westinghouse to pursue alternative designs on the same principles. He assigned to William Stanley the task of developing a device for commercial use in United States. Stanley's first patented design was for induction coils with single cores of soft iron and adjustable gaps to regulate the EMF present in the secondary winding (see image). This design was first used commercially in the US in 1886 but Westinghouse was intent on improving the Stanley design to make it (unlike the ZBD type) easy and cheap to produce.

Westinghouse, Stanley and associates soon developed an easier to manufacture core, consisting of a stack of thin 'Eshaped' iron plates, insulated by thin sheets of paper or other insulating material. Prewound copper coils could then be slid into place, and straight iron plates laid in to create a closed magnetic circuit. Westinghouse applied for a patent for the new low-cost design in December 1886; it was granted in July 1887.

REFERENCES

1. Knowlton, A.E. (Ed.) (1949). Standard Handbook for Electrical Engineers (8th ed.). McGraw-Hill. p. 597, Fig. 6–42.
2. Mack, James E.; Shoemaker, Thomas (2006). Chapter 15 - Distribution Transformers (PDF) (11th ed.). New York: McGraw-Hill. pp. 15–1 to 15–22. ISBN 0-07-146789-0.
3. Bedell, Frederick. "History of A-C Wave Form, Its Determination and Standardization". Transactions of the American Institute of Electrical Engineers. 61 (12): 864. doi:10.1109/T-AIEE.1942.5058456.

4. Brenner, Egon; Javid, Mansour (1959). "§18.1 'Symbols and Polarity of Mutual Inductance' in Chapter 18 – Circuits with Magnetic Circuits". Analysis of Electric Circuits. McGraw-Hill. pp. 598–599.
5. Brenner, pp. 598–600
6. Crosby, D. (1958). "The Ideal Transformer". IRE Transactions on Circuit.Theory. 5 (2): 145–145. doi:10.1109/TCT.1958.1086447.
7. Hameyer, Kay (2001). "§2.1.2 'Second Maxwell-Equation (Faraday's Law)' in Section 2 - Basics". Electrical s I: Basics, Design, Function, Operation. RWTH Aachen University Institute of Electrical s. pp. 11–12, eq. 2–12 to 2–15.
8. Heathcote, Martin (Nov 3, 1998). J & P Transformer Book (12th ed.). Newnes. pp. 2–3. ISBN 0-7506-1158-8.
9. Rajput, R.K. (2002). Alternating current s (3rd ed.). New Delhi: Laxmi Publications. p. 107. ISBN 9788170082224.
10. Calvert, James (2001). "Inside Transformers". University of Denver. Retrieved May 19, 2007.
11. Winders, John J., Jr. (2002). Power Transformer Principles and Applications. CRC. pp. 20–21.
12. Hameyer, Kay (2001). "§3.2 'Definition of Transformer Ratio' in Section 3 - Transformers". Electrical s I: Basics, Design, Function, Operation. RWTH Aachen University Institute of Electrical s. p. 27.
13. Knowlton, §6-13, p. 552
14. Miller, Wilhelm C.; Robbins, Allan H. (2013). Circuit analysis : theory and practice (5th ed.). Clifton Park, NY: Cengage Learning. p. 990. ISBN 978-1-1332-8100-9. Retrieved 25 September 2014.
15. Flanagan, pp. 1–2
16. Tcheslavski, Gleb V. (2008). "Slide 13 Impedance Transformation in Lecture 4: Transformers". ELEN 3441 Fundamentals of Power Engineering. Lamar University (TSU system member).
17. Parker, M. R; Ula, S.; Webb, W. E. (2005). "§2.5.5 'Transformers' & §10.1.3 'The Ideal Transformer'". In Whitaker, Jerry C. The Electronics Handbook (2nd ed.). Taylor & Francis. pp. 172, 1017. ISBN 0-8493-1889-0.
18. Kothari, D.P.; Nagrath, I.J. (2010). §3.7 'Transformer Testing' in Chapter 3 – Transformers (4th ed.). Tata McGraw-Hill. p. 73. ISBN 978-0-07-069967-0.
19. Brenner, pp. 589–590
20. "Polarity Markings on Instrument Transformers" (PDF). Retrieved 13 April 2013.
21. ANSI/IEEE C57.13, ANS Requirements for Instrument Transformers. New York, N.Y.: IEEE. 1978. p. 4 (§3.26). ISBN 0-7381-4299-9. (superseded, 1993)
22. "Connections - Polarity" (PDF). Retrieved 13 April 2013.
23. Say, M. G. (1984). Alternating Current s (5th ed.). Halsted Press. ISBN 0-470-27451-4.

24. ^ Jump up to:a b McLaren, P. G. (1984). Elementary Electric Power and s. pp. 68–74. ISBN 0-13-257601-5.
25. Say, p. 485
26. Terman, Frederick E. (1955). Electronic and Radio Engineering (4th ed.). New York: McGraw-Hill. p. 15.
27. Heathcote, p. 4
28. Knowlton, pp. 585-586
29. Hameyer, p. 39
30. Daniels, A. R. Introduction to Electrical s. pp. 47–49.
31. Say, pp. 142-143
32. IEC Std 60404-2 Magnetic Materials – Part 2: Methods of Measurement of the Magnetic Properties . . .
33. Hameyer, p. 11, eq. 2-13
34. Billings, Keith (1999). Switchmode Power Supply Handbook. McGraw-Hill. ISBN 0-07-006719-8.
35. "400 Hz Electrical Systems". Aerospaceweb.org. Retrieved May 21, 2007.
36. Gururaj, B.I. (June 1963). "Natural Frequencies of 3-Phase Transformer Windings". IEEE Transactions on Power Apparatus and Systems. 82 (66): 318–329. doi:10.1109/TPAS.1963.291359.
37. De Keulenaer, Hans; Chapman, David; Fassbinder, Stefan; McDermott, Mike (2001). "The Scope for Energy Saving in the EU through the Use of Energy-Efficient Electricity Distribution Transformers" (PDF). Institution of Engineering and Technology. Retrieved 10 July 2014.
38. Kubo, T.; Sachs, H.; Nadel, S. (2001). Opportunities for New Appliance and Equipment Efficiency Standards. American Council for an Energy-Efficient Economy. p. 39, fig. 1. Retrieved June 21, 2009.
39. Riemersma, H.; Eckels, P.; Barton, M.; Murphy, J.; Litz, D.; Roach, J. (1981). "Application of Superconducting Technology to Power Transformers". IEEE Transactions on Power Apparatus and Systems. PAS-100 (7): 3398. doi:10.1109/TPAS.1981.316682.
40. Heathcote, pp. 41-42
41. Knowlton, pp. 49 (§2.67) & 323 (§4.279)
42. EE-Reviewonline.com. "Steinmetz's Formula for Magnetic Hysteresis". Retrieved 7 February 2013.
43. "Understanding Transformer Noise" (PDF). FP. Retrieved 30 January 2013.
44. Nailen, Richard (May 2005). "Why We Must Be Concerned With Transformers". Electrical Apparatus.
45. Pansini, Anthony J. (1999). Electrical Transformers and Power Equipment. Fairmont Press. p. 23. ISBN 0-88173-311-3.

46. Del Vecchio, Robert M.; et al. (2002). Transformer Design Principles: With Applications to Core-Form Power Transformers. Boca Raton: CRC Press. pp. 10–11, Fig. 1.8. ISBN 90-5699-703-3.

47. Knowlton, p. 562

48. Hydroelectric Research and Technical Services Group. "Transformers: Basics, Maintenance, and Diagnostics"(PDF). U.S. Dept. of the Interior, Bureau of Reclamation. p. 12. Retrieved Mar 27, 2012.

49. EM 1110-2-3006 (1994). "Chapter 4 - Power Transformers" (PDF). Engineering and Design – Hydroelectric Power Plants Electrical Design. U.S. Army Corps of Engineers. p. 4-1.

50. Hindmarsh, J. (1984). Electrical s and Their Applications. Oxford: Pergamon Press. pp. 29–31. ISBN 0-08-030573-3.

51. Gottlieb, Irving (1998). Practical Transformer Handbook. Newnes. p. 4. ISBN 0-7506-3992-X.

52. Allan, D.J. (Jan 1991). "Power Transformers – The Second Century". Power Engineering Journal. 5 (1): 5–14. doi:10.1049/pe:19910004.

53. Kulkarni, S. V.; Khaparde, S. A. (May 24, 2004). Transformer Engineering: Design and Practice. CRC. pp. 36–37. ISBN 0-8247-5653-3.

54. Hindmarsh, pp. 29-31

55. McLyman, Colonel Wm. T. (2004). Transformer and Inductor Design Handbook. CRC. Chap. 3, pp. 9–14. ISBN 0-8247-5393-3.

56. Sim, H. Jim; Digby; Scott H. (2004). "§2.1 Power Transformers in Chap. 2 - Equipment Types". In L. L. Grigsby; James H. Harlow. The Electric Power Engineering Handbook (Online ed.). CRC Press. §2.1.7 & §2.1.6.2.1. ISBN 0-8493-8578-4.

57. Boteler, D. H.; Pirjola, R. J.; Nevanlinna, H. (1998). "The Effects of Geomagnetic Disturbances On Electrical Systems at the Earth's Surface". Advances in Space Research. 22: 17–27. doi:10.1016/S0273-1177(97)01096-X.

58. Hasegawa, Ryusuke (June 2, 2000). "Present Status of Amorphous Soft Magnetic Alloys". Journal of Magnetism and Magnetic Materials. 215 216: 240–245. doi:10.1016/S0304-8853(00)00126-8.

59. McLyman, Chap. 3 p. 1

60. "Toroidal Line Power Transformers. Power Ratings Tripled. | Magnetics Magazine". www.magneticsmagazine.com. Retrieved 2016-09-23.

61. Lee, Reuben. "Air-Core Transformers". Electronic Transformers and Circuits. Retrieved May 22, 2007.

62. Dixon, L.H., Jr. (1997). "Eddy Current Losses in Transformer Windings" (PDF). Texas Instrument: R2–1–to–R2–10.

63. Central Electricity Generating Board (1982). Modern Power Station Practice. Pergamon Press.

64. Dixon, Lloyd (2001). "Power Transformer Design" (PDF). Magnetics Design Handbook. Texas Instruments.
65. Lane, Keith (2007). "The Basics of Large Dry-Type Transformers". EC&M. Retrieved 29 January 2013.
66. Heathcote, pp. 720–723
67. Tillman, Robert F. (2004). "§3.4 Load and Thermal Performance in Chap. 3 - Ancillary Topics" (PDF). In James H. Harlow (ed.); L. L. Grigsby (Series ed.). The Electric Power Engineering Handbook (Online ed.). CRC Press. p. §3.4.8. ISBN 0-8493-8578-4.
68. Walling, Reigh; Shattuck; G. Bruce (May 2007). Distribution Transformer Thermal Behaviour and Aging in Local-Delivery Distribution Systems (PDF). 19th International Conference on Electricity Distribution. Paper 0720. Retrieved 11 February 2013.
69. Kimberly, E.E. "Permissible Temperatures for Insulation". Retrieved 12 February 2013.
70. Pansini, p. 32
71. Willis, H. Lee (2004). Power Distribution Planning Reference Book. CRC Press. p. 403. ISBN 0-8247-4875-1.
72. Hartley, William H. (2003). Analysis of Transformer Failures. 36th Annual Conference of the International Association of Engineering Insurers. p. 7 (fig. 6). Retrieved 30 January 2013.
73. Hartley, William H. (~2011). "An Analysis of Transformer Failures, Part 1 – 1988 through 1997". The Locomotive. Retrieved 30 January 2013.
74. Prevost, Thomas A.; et al. (Nov 2006). "Estimation of Insulation Life Based on a Dual Temperature Aging Model"(PDF). Weidmann. p. 1. Retrieved Mar 30, 2012.
75. Sen, P.K.; et al. (2011). "PSERC Pub. 11-02 Transformer Overloading and Assessment of Loss-of-Life for Liquid-Filled Transformers" (PDF). Power Systems Engineering Research Center, Arizona State University. Retrieved 11 January 2013.
76. "ASTDR ToxFAQs for Polychlorinated Biphenyls". 2001. Retrieved June 10, 2007.
77. Kulkarni, pp. 2–3
78. AFBI (2011). "9. Contaminants" (PDF). State of the Seas Report. Agri-Food and Biosciences Institute & Northern Ireland Environment Agency. p. 71. ISBN 978-1-907053-20-7. Unknown ID 9977.
79. McDonald, C. J.; Tourangeau, R. E. (1986). PCBs, Question and Answer Guide Concerning Polychlorinated Biphenyls(PDF). Government of Canada: Environment Canada Department. ISBN 0-662-14595-X. Retrieved Nov 7, 2007.
80. Mehta, S.P.; Aversa, N.; Walker, M.S. (Jul 1997). "Transforming Transformers [Superconducting windings]"(PDF). IEEE Spectrum. 34 (7): 43–49. doi:10.1109/6.609815. Retrieved 14 November 2012.

81. Pansini, pp. 66–67

82. Fink, Donald G.; Beatty, H. Wayne (Eds.) (1978). Standard Handbook for Electrical Engineers (11th ed.). McGraw Hill. pp. 10–38 through 10–40. ISBN 978-0-07-020974-9.

83. Figueroa, Elisa; et al. (Jan–Feb 2009). "Low Frequency Heating Field Dry-Out of a 750 MVA 500 kV Auto Transformer" (PDF). Electricity Today. Retrieved Feb 28, 2012.

84. Ryan, Hugh M. (2001). High Voltage Engineering and Testing. Institution Electrical Engineers. pp. 416–417. ISBN 0-85296-775-6.

85. Lawhead, Larry; Hamilton, Randy; Horak, John (2006). "Three Phase Transformer Winding Configurations and Differential Relay Compensation" (PDF). Georgia Tech 60th Protective Relay Conference. pp. 8–10. Retrieved Feb 23, 2012.

86. Beeman, Donald (Ed.) (1955). Industrial Power Systems Handbook. McGraw-Hill. pp. 349–364.

87. Brown, BIll. "Section 6 Grounding Systems" (PDF). Schneider. pp. 9–12. Retrieved 18 January 2013.

88. Beltz, Robert; Peacock, Ian; Vilcheck, William (2000). Application Considerations for High Resistance Ground Retrofits in Pulp and Paper Mills. Pulp and Paper Industry Technical Conference. pp. 33–40. doi:10.1109/PAPCON.2000.854186.

89. Knowlton, §6–7, pp. 549–550

90. "Power Transformers On Triad Magnetics". catalog.triadmagnetics.com. Retrieved 2016-09-23.

91. IEEE PES TC (Fall 2011). "Discussion of Class I & II Terminology" (PDF). IEEE PES Transformer Committee. p. slide 6. Retrieved 27 January 2013.

92. Knowlton, §12-341, fig. 12-95, p. 1085

93. AEMO; et al. (2012). "Joint Consultation Paper - Western Metropolitan Melbourne Transmission Connection and Subtransmission Capacity". p. 11. Archived from the original (PDF) on 4 January 2013.

94. Heathcote, p. 1

95. "How the Electricity Grid Works". Retrieved 2016-09-23.

96. Poyser, Arthur William (1892). Magnetism and Electricity: A Manual for Students in Advanced Classes. London and New York: Longmans, Green, & Co. p. 285, fig. 248.

97. "A Brief History of Electromagnetism" (PDF).

98. "Electromagnetism". Smithsonian Institution Archives.

99. http://www.ryancmacpherson.com/publications/3-book-reviews/49-joseph-henry-the-rise-of-an-american-scientist.html

100. Guarnieri, M. (2013). "Who Invented the Transformer?". IEEE Industrial Electronics Magazine. 7 (4): 56–59. doi:10.1109/MIE.2013.2283834.

101. Chow, Tai L. (2006). Introduction to Electromagnetic Theory: A Modern

Perspective. Sudbury, Mass.: Jones and Bartlett Publishers. p. 171. ISBN 0-7637-3827-1.

102. Faraday, Michael (1834). "Experimental Researches on Electricity, 7th Series". Philosophical Transactions of the Royal Society. 124: 77–122. doi:10.1098/rstl.1834.0008.

103. French patent No.115793 granted to Jablochkoff on November 30, 1876 with the title "Disposition de courants, destinée à l'éclairage par la lumière électrique», reference to it is available in the following book: Subject-Matter Index of Patents for Inventions (Brevets D'Invention) Granted in France from 1791 to 1876 Inclusive. Washington. 1883. p. 248.

104. "Stanley Transformer". Los Alamos National Laboratory; University of Florida. Retrieved Jan 9, 2009.

105. De Fonveille, W. (Jan 22, 1880). "Gas and Electricity in Paris". Nature. 21 (534): 283. Bibcode:1880Natur..21..282D. doi:10.1038/021282b0. Retrieved Jan 9, 2009.

106. Hughes, Thomas P. (1993). Networks of Power: Electrification in Western Society, 1880-1930. Baltimore: The Johns Hopkins University Press. p. 96. ISBN 0-8018-2873-2. Retrieved Sep 9, 2009.

107. Uppenborn, F. J. (1889). History of the Transformer. London: E. & F. N. Spon. pp. 35–41.

108. Coltman, J. W. (Jan 1988). "The Transformer". Scientific American. pp. 86–95. OSTI 6851152.

109. Stanley, William, Jr. "Induction Coil". U.S. Patent 349 311, issued Sept. 21, 1886. Retrieved July 13, 2009.

110. Hughes, p. 95

111. Károly, Simonyi. "The Faraday Law With a Magnetic Ohm's Law". Természet Világa. Retrieved Mar 1, 2012.

112. Lucas, J.R. "Historical Development of the Transformer"(PDF). IEE Sri Lanka Centre. Retrieved Mar 1, 2012.

113. Halacsy, A. A.; Von Fuchs, G. H. (April 1961). "Transformer Invented 75 Years Ago". IEEE Transactions of the American Institute of Electrical Engineers. 80 (3): 121–125. doi:10.1109/AIEEPAS.1961.4500994. Retrieved Feb 29, 2012.

114. Jeszenszky, Sándor. "Electrostatics and Electrodynamics at Pest University in the Mid-19th Century" (PDF). University of Pavia. Retrieved Mar 3, 2012.

115. "Hungarian Inventors and Their Inventions". Institute for Developing Alternative Energy in Latin America. Retrieved Mar 3, 2012.

116. "Bláthy, Ottó Titusz". Budapest University of Technology and Economics, National Technical Information Centre and Library. Retrieved Feb 29, 2012.

117. "Bláthy, Ottó Titusz (1860–1939)". Hungarian Patent Office. Retrieved Jan 29, 2004.

118. Zipernowsky, K.; Déri, M.; Bláthy, O.T. "Induction Coil"(PDF). U.S. Patent 352 105, issued Nov. 2, 1886. Retrieved July 8, 2009.

119. Smil, Vaclav (2005). Creating the Twentieth Century: Technical Innovations of 1867—1914 and Their Lasting Impact. Oxford: Oxford University Press. p. 71. ISBN 978-0-19-803774-3.

120. Nagy, Árpád Zoltán (Oct 11, 1996). "Lecture to Mark the 100th Anniversary of the Discovery of the Electron in 1897 (preliminary text)". Budapest. Retrieved July 9, 2009.

121. Oxford English Dictionary (2nd ed.). Oxford University Press. 1989.

122. Hospitalier, Édouard (1882). The Modern Applications of Electricity. Translated by Julius Maier. New York: D. Appleton & Co. p. 103.

123. "Ottó Bláthy, Miksa Déri, Károly Zipernowsky". IEC Techline. Retrieved Apr 16, 2010.

124. Skrabec, Quentin R. (2007). George Westinghouse: Gentle Genius. Algora Publishing. p. 102. ISBN 978-0-87586-508-9.

125. Coltman, J.W. (Jan–Feb 2002). "The Transformer [Historical Overview]". Industry Applications Magazine, IEEE. 8 (1): 8–15. doi:10.1109/2943.974352. Retrieved Feb 29, 2012.

126. International Electrotechnical Commission. Otto Blathy, Miksa Déri, Károly Zipernowsky. IEC History. Retrieved May 17, 2007.

127. Westinghouse, George, Jr. "Electrical Converter". U.S. Patent 366 362, issued July 12, 1887.

128. Neidhöfer, Gerhard (2008). Michael von Dolivo-Dobrowolsky and Three-Phase: The Beginnings of Modern e Technology and Power Supply (in German). In collaboration with VDE "History of Electrical Engineering" Committee (2 ed.). Berlin: VDE-Verl. ISBN 978-3-8007-3115-2.

129. Uth, Robert (Dec 12, 2000). "Tesla Coil". Tesla: Master of Lightning. PBS.org. Retrieved May 20, 2008.

130. Tesla, Nikola. "System of Electrical Lighting". U.S. Patent 454 622, issued June 23, 1891.

CHAPTER 7

Switchgear And Protection

Chapter Contents

- **Introduction**
- **Components**
- **Functions**

The apparatus used for switching, controlling and protecting the electrical circuits and equipment is known as switchgear.

INTRODUCTION

In an electric power system, switchgear is the combination of electrical disconnect switches, fuses or circuit breakers used to control, protect and isolate electrical equipment. Switchgear is used both to de-energize equipment to allow work to be done and to clear faults downstream. This type of equipment is directly linked to the reliability of the electricity supply.

The earliest central power stations used simple open knife switches, mounted on insulating panels of marble or asbestos. Power levels and voltages rapidly escalated, making opening manually operated switches too dangerous for anything other than isolation of a de-energized circuit. Oil-filled equipment allowed arc energy to be contained and safely controlled. By the early 20th century, a switchgear line-up would be a metal-enclosed structure with electrically operated switching elements, using oil circuit breakers. Today, oil-filled equipment has largely been replaced by finger-blast, vacuum, or SF6 equipment, allowing large currents and power levels to be safely controlled by automatic equipment.

High-voltage switchgear was invented at the end of the 19th century for operating motors and other electric machines. The technology has been improved over time and can now be used with voltages up to 1,100 kV.

Typically, switchgears in substations are located on both the high- and low-voltage sides of large power transformers. The switchgear on the low-voltage side of the transformers may be located in a building, with medium-voltage circuit breakers for distribution circuits, along with metering, control, and protection equipment. For industrial applications, a transformer and switchgear line-up may be combined in one housing, called a unitized substation (USS).

High voltage switchgear

A section of a large switchgear panel, in this case, used to control on-board casino boat power generation.

Tram switchgear

COMPONENTS

A switchgear has two types of components:

- Power conducting components, such as switches, circuit breakers, fuses, and lightning arrestors, that conduct or interrupt the flow of electrical power
- Control systems such as control panels, current transformers, potential transformers, protective relays, and associated circuitry, that monitor, control, and protect the power conducting components

FUNCTIONS

One of the basic functions of switchgear is protection, which is interruption of short-circuit and overload fault currents while maintaining service to unaffected circuits. Switchgear also provides isolation of circuits from power supplies. Switchgear is also used to enhance system availability by allowing more than one source to feed a load.

This circuit breaker uses both SF6 and air as insulation.

Early switchgear (about 1910)

Switchgears are as old as electricity generation. The first models were very primitive: all components were simply fixed to a wall. Later they were mounted on wooden panels. For reasons of fire protection, the wood was replaced by slate or marble. This led to a further improvement, because the switching and measuring devices could be attached to the front, while the wiring was on the back.

Housing

Switchgear for lower voltages may be entirely enclosed within a building. For higher voltages (over about 66 kV), switchgear is typically mounted outdoors and insulated by air, although this requires a large amount of space. Gas-insulated switchgear saves space compared with air-insulated equipment, although the equipment cost is higher. Oil insulated switchgear presents an oil spill hazard.

Switches may be manually operated or have motor drives to allow for remote control.

Circuit Breaker Types

A switchgear may be a simple open-air isolator switch or it may be insulated by some other substance. An effective although more costly form of switchgear is the gas-insulated switchgear (GIS), where the conductors and contacts are insulated by pressurized sulfur hexafluoride gas (SF_6). Other common types are oil or vacuum insulated switchgear.

The combination of equipment within the switchgear enclosure allows them to interrupt fault currents of thousands of amps. A circuit breaker (within a switchgear enclosure) is the primary component that interrupts fault currents. The quenching of the arc when the circuit breaker pulls apart the contacts (disconnects the circuit) requires careful design. Circuit breakers fall into these five types:

Oil

Oil circuit breakers rely upon vaporization of some of the oil to blast a jet of oil along the path of the arc. The vapor released by the arcing consists of hydrogen gas. Mineral oil has better insulating property than air. Whenever there is a separation of current carrying contacts in the oil, the arc in circuit breaker is initialized at the moment of separation of contacts, and due to this arc the oil is vaporized and decomposed in mostly hydrogen gas and ultimately creates a hydrogen bubble around the electric arc. This highly compressed gas bubble around the arc prevents re-striking of the arc after current reaches zero crossing of the cycle. The oil circuit breaker is one of the oldest type of circuit breakers.

Air

Air circuit breakers may use compressed air (puff) or the magnetic force of the arc itself to elongate the arc. As the length of the sustainable arc is dependent on the available voltage, the elongated arc will eventually exhaust itself. Alternatively, the contacts are rapidly swung into a small sealed chamber, the escaping of the displaced air thus blowing out the arc.

Circuit breakers are usually able to terminate all current flow very quickly: typically between 30 ms and 150 ms depending upon the age and construction of the device.

Gas

Gas (SF_6) circuit breakers sometimes stretch the arc using a magnetic field, and then rely upon the dielectric strength of the SF_6 gas to quench the stretched arc.

Hybrid

Hybrid switchgear is a type which combines the components of traditional air-insulated switchgear (AIS) and SF_6 gas-insulated switchgear (GIS) technologies. It is characterized by a compact and modular design, which encompasses several different functions in one module.

Vacuum

Circuit breakers with vacuum interrupters have minimal arcing characteristics (as there is nothing to ionize other than the contact material), so the arc quenches when it is stretched by a small amount (<2–8 mm). Near zero current the arc is not hot enough to maintain a plasma, and current ceases; the gap can then withstand the rise of voltage. Vacuum circuit breakers are frequently used in modern medium-voltage switchgear to 40,500 volts. Unlike the other types, they are inherently unsuitable for interrupting DC faults. The reason vacuum circuit breakers are unsuitable for breaking high DC voltages is that with DC there is no "current zero" period. The plasma arc can feed itself by continuing to gasify the contact material.

Carbon Dioxide (CO_2)

Breakers that use carbon dioxide as the insulating and arc extinguishing medium work on the same principles as a sulfur hexafluoride (SF_6) breaker. Because SF_6 is a greenhouse gas more potent than CO_2, by switching from SF_6 to CO_2 it is possible to reduce the greenhouse gas emissions by 10 tons during the product lifecycle.

Protective Circuitry

Circuit Breakers and Fuses

Circuit breakers and fuses disconnect when current exceeds a predetermined safe level. However they cannot sense other critical faults, such as unbalanced currents—for example, when a transformer winding contacts ground. By themselves, circuit breakers and fuses cannot distinguish between short circuits and high levels of electrical demand.

Merz-Price Circulating Current Scheme

Differential protection depends upon Kirchhoff's current law, which states that the sum of currents entering or leaving a circuit node must equal zero. Using this principle to implement differential protection, any section of a conductive path may be considered a node, The conductive path could be a transmission line, a winding of a transformer, a winding in a motor, or a winding in the stator of an alternator. This form of protection works best when both ends of the conductive path are physically close to each other. This scheme was invented in Great Britain by Charles Hesterman Merz and Bernard Price.

Two identical current transformers are used for each winding of a transformer, stator, or other device. The current transformers are placed around opposite ends of a winding. The current through both ends should be identical. A protective relay detects any imbalance in currents, and trips circuit breakers to isolate the device. In the case of a transformer, the circuit breakers on both the primary and secondary would open.

Distance Relays

A short circuit at the end of a long transmission line appears similar to a normal load, because of the impedance of the transmission line limits the fault current. A distance relay detects a fault by comparing the voltage and current on the transmission line. A large current along with a voltage drop indicates a fault.

Classification

Several different classifications of switchgear can be made:

- By the current rating.
- By interrupting rating (maximum short circuit current kAIC that the device can safely interrupt)

    - Circuit breakers can open and close on fault currents
    - Load-break/Load-make switches can switch normal system load currents
    - Isolators are off load disconnectors which are to be operated after Circuit Breakers, or else if the load current is very small
- By voltage class:
    - Low voltage (less than 1 kV AC)
    - Medium voltage (1 kV AC through to approximately 75 kV AC)
    - High voltage (75 kV to about 230 kV AC)
    - Extra high voltage, ultra high voltage (more than 230 kV)
- By insulating medium:
    - Air
    - Gas (SF_6 or mixtures)
    - Oil
    - Vacuum
    - Carbon dioxide (CO_2)
- By construction type:
    - Indoor (further classified by IP (Ingress Protection) class or NEMA enclosure type)
    - Outdoor
    - Industrial
    - Utility
    - Marine
    - Draw-out elements (removable without many tools)
    - Fixed elements (bolted fasteners)
    - Live-front
    - Dead-front
    - Open
    - Metal-enclosed (ME) — A switchgear assembly completely enclosed on all sides and the top with sheet metal.
    - Metal-clad (MC) — A more expensive variety of metal-enclosed switchgear that has the following characteristics: the main switching and interrupting device of removable type; grounded metal barriers to separate compartments and enclose all major circuits and parts; mechanical interlocks; insulated bus conductors and other features.
    - Arc-resistant
- By IEC degree of internal separation
    - No Separation (Form 1)
    - Busbars separated from functional units (Form 2a, 2b, 3a, 3b, 4a, 4b)

- Terminals for external conductors separated from busbars (Form 2b, 3b, 4a, 4b)
- Terminals for external conductors separated from functional units but not from each other (Form 3a, 3b)
- Functional units separated from each other (Form 3a, 3b, 4a, 4b)
- Terminals for external conductors separated from each other (Form 4a, 4b)
- Terminals for external conductors separate from their associated functional unit (Form 4b)

■ By interrupting device:
- Fuses
- Air Circuit Breaker
- Minimum Oil Circuit Breaker
- Oil Circuit Breaker
- Vacuum Circuit Breaker
- Gas (SF_6) Circuit breaker
- CO_2 Circuit Breaker

■ By operating method:
- Manually operated
- Motor/stored energy operated
- Solenoid operated

■ By type of current:
- Alternating current
- Direct current

■ By application:
- Transmission system
- Distribution

■ By purpose
- Isolating switches (disconnectors)
- Load-break switches.
- Grounding (earthing) switches

A single line-up may incorporate several different types of devices, for example, air-insulated bus, vacuum circuit breakers, and manually operated switches may all exist in the same row of cubicles.

Ratings, design, specifications and details of switchgear are set by a multitude of standards. In North America mostly IEEE and ANSI standards are used, much of the rest of the world uses IEC standards, sometimes with local national derivatives or variations.

Safety

To help ensure safe operation sequences of switchgear, trapped key interlocking provides predefined scenarios of operation. For example, if only one of two sources of supply are permitted to be connected at a given time, the interlock scheme may require that the first switch must be opened to release a key that will allow closing the second switch. Complex schemes are possible.

Indoor switchgear can also be type tested for internal arc containment (e.g., IEC 62271-200). This test is important for user safety as modern switchgear is capable of switching large currents.

Switchgear is often inspected using thermal imaging to assess the state of the system and predict failures before they occur. Other methods include partial discharge (PD) testing, using either fixed or portable testers, and acoustic emission testing using surface-mounted transducers (for oil equipment) or ultrasonic detectors used in outdoor switchyards. Temperature sensors fitted to cables to the switchgear can permanently monitor temperature build-up. SF_6 equipment is invariably fitted with alarms and interlocks to warn of loss of pressure, and to prevent operation if the pressure falls too low.

The increasing awareness of dangers associated with high fault levels has resulted in network operators specifying closed-door operations for earth switches and racking breakers. Many European power companies have banned operators from switch rooms while operating. Remote racking systems are available which allow an operator to rack switchgear from a remote location without the need to wear a protective arc flash hazard suit.

245 kV circuit breaker in air insulated substation 420 kV gas insulated switchgear

REFERENCES

1. British Pattern GB 20069 Improvements in Apparatus for Controlling the Application or Use of Electric Currents of High Tension and Great Quantity in 1893, on espacenet.com
2. Lin Jiming et al., Transient characteristics of 1 100 kV circuit-breakers, International Symposium on International Standards for Ultra High Voltage, Beijing, Juillet 2007.
3. (German) Allgemeine Elektricitäts-Gesellschaft (ed) AEG Hilfsbuch für elektrische Licht- und Kraftanlagen 6th Ed., W. Girardet, Essen 1953
4. http://www.electrical4u.com/oil-circuit-breaker-bulk-and-minimum-oil-circuit-breaker/
5. "Switzerland : ABB breaks new ground with environment friendly high-voltage circuit breaker.". Retrieved 9 July 2013.
6. Robert Monro Black (January 1983). The History of Electric Wires and Cables. IET. pp. 101–. ISBN 978-0-86341-001-7.
7. Robert W. Smeaton (ed) Switchgear and Control Handbook 3rd Ed., McGraw Hill, New York 1997 ISBN 0-07-058451-6
8. IEEE Std C37.20.2-1999. IEEE Standard for Metal-Clad Switchgear.
9. IEEE Std C37.100-1992. IEEE Standard Definitions for Power Switchgear.
10. "Metal-Clad vs Metal-Enclosed". ELECTRICAL ENGINEERS AND MASTER ELECTRICIANS PORTAL. November 4, 2008. Retrieved June 28, 2016.
11. IEC Standard EN 60439 part 1 Table 6A
12. (French) Norme CEI 60265-1 Interrupteurs pour tension assignée supérieure à 1 kV et inférieure à 52 kV ArchivedSeptember 30, 2007, at the Wayback Machine.

13. (French) Norme CEI 60265-2 Interrupteurs pour tension assignée supérieure à 52 kV Archived September 30, 2007, at the Wayback Machine.

14. https://www.energy.siemens.com/cms/00000013/aune/Documents/Medium%20Voltage%20Arc%20Fault%20Containment.pdfArchived March 18, 2009, at the Wayback Machine.

CHAPTER 8

Electrical Power Transmission

Chapter Contents

Electrical power is generated at different generating stations. These generating stations are not necessarily situated at the load center. During construction of generating station number of factors are to be considered from economical point of view. These all factors may not be easily available at load center; hence generating stations are not normally situated very nearer to load center. Load center is the place where maximum power is consumed. Hence there must be some means by which the generated power must be transmitted to the load center. Electrical transmission system is the means of transmitting power from generating station to different load centers.

INTRODUCTION

Electric power transmission is the bulk movement of electrical energy from a generating site, such as a power plant, to an electrical substation. The interconnected lines which facilitate this movement are known as a transmission network. This is distinct from the local wiring between high-voltage substations and customers, which is typically referred to as electric power distribution. The combined transmission and distribution network is known as the "power grid" in North America, or just "the grid". In the United Kingdom, the network is known as the "National Grid".

A wide area synchronous grid, also known as an "interconnection" in North America, directly connects a large number of generators delivering AC power with the same relative *frequency* to a large number of consumers. For example, there are four major interconnections in North America (the Western Interconnection, the Eastern Interconnection, the Quebec Interconnection and the Electric Reliability Council of Texas (ERCOT) grid), and one large grid for most of continental Europe.

Historically, transmission and distribution lines were owned by the same company, but starting in the 1990s, many countries have liberalized the regulation of the electricity market in ways that have led to the separation of the electricity transmission business from the distribution business.

Two-circuit, single-voltage power transmission line; "Bundled" 4-ways

Power lines in the San Francisco Bay Area

Power lines in Qatar

SYSTEM

Most transmission lines are high-voltage three-phase alternating current (AC), although single phase AC is sometimes used in railway electrification systems. High-voltage direct-current (HVDC) technology is used for greater efficiency over very long distances (typically hundreds of miles). HVDC technology is also used in submarine power cables (typically longer than 30 miles (50 km)), and in the interchange of power between grids that are not mutually synchronized. HVDC links are used to stabilize large power distribution networks where sudden new loads, or blackouts, in one part of a network can result in synchronization problems and cascading failures.

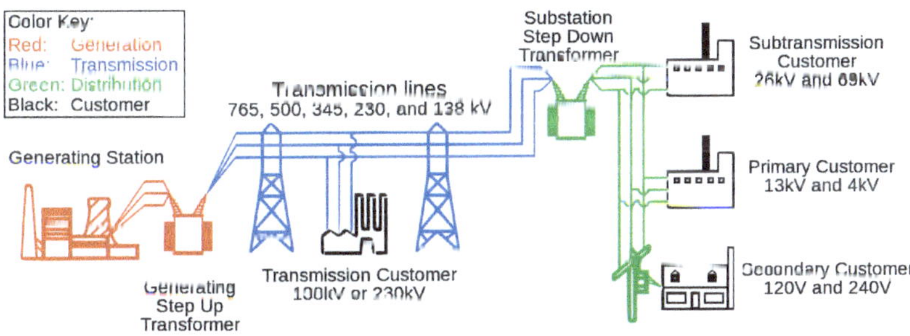

Diagram of an electric power system; transmission system is in blue

Electricity is transmitted at high voltages (115 kV or above) to reduce the energy loss which occurs in long-distance transmission. Power is usually transmitted through overhead power lines. Underground power transmission has a significantly higher installation cost and greater operational limitations, but reduced maintenance costs. Underground transmission is sometimes used in urban areas or environmentally sensitive locations.

A lack of electrical energy storage facilities in transmission systems leads to a key limitation. Electrical energy must be generated at the same rate at which it is consumed. A sophisticated control system is required to ensure that the power generation very closely matches the demand. If the demand for power exceeds supply, the imbalance can cause generation plant(s) and transmission equipment to automatically disconnect and/or shut down to prevent damage. In the worst case, this may lead to a cascading series of shut downs and a major regional blackout. Examples include the US Northeast blackouts of 1965, 1977, 2003, and major blackouts in other US regions in 1996 and 2011. Electric transmission networks are interconnected into regional, national, and even continent wide networks to reduce the risk of such a failure by providing multiple redundant, alternative routes for power to flow should such shut downs occur. Transmission companies determine the maximum reliable capacity of each line (ordinarily less than its physical or thermal limit) to ensure that spare capacity is available in the event of a failure in another part of the network.

OVERHEAD TRANSMISSION

3-phase high-voltage lines in Washington State, "Bundled" 3-ways

Four-circuit, two-voltage power transmission line; "Bundled" 2-ways

A typical ACSR. The conductor consists of seven strands of steel surrounded by four layers of aluminium.

High-voltage overhead conductors are not covered by insulation. The conductor material is nearly always an aluminum alloy, made into several strands and possibly reinforced with steel strands. Copper was sometimes used for overhead transmission, but aluminum is lighter, yields only marginally reduced performance and costs much less. Overhead conductors are a commodity supplied by several companies worldwide. Improved conductor material and shapes are regularly used to allow increased capacity and modernize transmission circuits. Conductor sizes range from 12 mm^2 (#6 American wire gauge) to 750 mm^2 (1,590,000 circular mils area), with varying resistance and current-carrying capacity. Thicker wires would lead to a relatively small increase in capacity due to the skin effect, that causes most of the current to flow close to the surface of the wire. Because of this current limitation, multiple parallel cables (called bundle conductors) are used when higher capacity is needed. Bundle conductors are also used at high voltages to reduce energy loss caused by corona discharge.

Today, transmission-level voltages are usually considered to be 110 kV and above. Lower voltages, such as 66 kV and 33 kV, are usually considered subtransmission voltages, but are occasionally used on long lines with light loads. Voltages less than 33 kV are usually used for distribution. Voltages above 765 kV are considered extra high voltage and require different designs compared to equipment used at lower voltages.

Since overhead transmission wires depend on air for insulation, the design of these lines requires minimum clearances to be observed to maintain safety. Adverse weather conditions, such as high wind and low temperatures, can lead to power outages. Wind speeds as low as 23 knots (43 km/h) can permit conductors to encroach operating clearances, resulting in a flashover and loss of supply. Oscillatory motion of the physical line can be termed gallop or flutter depending on the frequency and amplitude of oscillation.

Underground Transmission

Electric power can also be transmitted by underground power cables instead of overhead power lines. Underground cables take up less right-of-way than overhead lines, have lower visibility, and are less affected by bad weather. However, costs of insulated cable and excavation are much higher than overhead construction. Faults in buried transmission lines take longer to locate and repair. Underground lines are strictly limited by their thermal capacity, which permits less overload or re-rating than overhead lines. Long underground AC cables have significant capacitance, which may reduce their ability to provide useful power to loads beyond 50 mi (80 km). Long underground DC cables have no such issue and can run for thousands of miles.

New York City streets in 1890. Besides telegraph lines, multiple electric lines were required for each class of device requiring different voltages

In the early days of commercial electric power, transmission of electric power at the same voltage as used by lighting and mechanical loads restricted the distance between generating plant and consumers. In 1882, generation was with direct current (DC), which could not easily be increased in voltage for long-distance transmission. Different classes of loads (for example, lighting, fixed motors, and traction/railway systems) required different voltages, and so used different generators and circuits.

Due to this specialization of lines and because transmission was inefficient for low-voltage high-current circuits, generators needed to be near their loads. It seemed, at the time, that the industry would develop into what is now known as a distributed generation system with large numbers of small generators located near their loads.

The transmission of electric power with alternating current (AC) became possible after Lucien Gaulard and John Dixon Gibbs built what they called the secondary generator, an early transformer provided with 1:1 turn ratio and open magnetic circuit, in 1881.

The first long distance (34 km (21 mi)) AC line was built for the 1884 International Exhibition of Turin, Italy. It was powered by a 2000-V, 130 Hz Siemens & Halske alternator and featured several Gaulard secondary generators with their primary windings connected in series, which fed incandescent lamps. The system proved the feasibility of AC electric power transmission on long distances.

A very first operative AC line was put into service in 1885 in via dei Cerchi, Rome, Italy, for public lighting. It was powered by two Siemens & Halske alternators rated 30 hp (22 kW), 2000-V at 120 Hz and used 19 km of cables and 200 parallel-connected 2000-V/20-V step-down transformers provided with a closed magnetic circuit, one for each lamp. Few months later it was followed by the first British AC system, which was put into service at the Grosvenor Gallery, London. It also featured Siemens alternators and 2400-V/100-V step-down transformers, one per user, with shunt-connected primaries.

Working for Westinghouse, William Stanley Jr. spent his time recovering from illness in Great Barrington installing what is considered the worlds first practical AC transformer system.

Working from what he considered an impractical Gaulard-Gibbs design, electrical engineer William Stanley, Jr. developed what is considered the first practical series AC transformer in 1885. Working with the support of George Westinghouse, in 1886 he installed demonstration transformer based alternating current lighting system in Great Barrington, Massachusetts. Powered by a steam engine driven 500-V Siemens generator, voltage was stepped down to 100 volts using the new Stanley transformer to power incandescent lamps at 23 businesses along main street with very little power loss over 4000 feet. This practical demonstration of a transformer/alternating current lighting system would lead Westinghouse to begin installing AC based systems later that year.

In 1888 alternating current systems gained further viability with introduction of a functional AC motor, something these systems had lacked up till then. The design, an induction motor running on polyphase current, was independently invented by Galileo Ferraris and Nikola Tesla (with Tesla's design being licensed by Westinghouse in the US). This design was further developed into the modern practical three-phase form by Mikhail Dolivo-Dobrovolsky and Charles Eugene Lancelot Brown.

The late 1880s/early 1890s would see a financial merger of many smaller electric companies into a few larger corporations such as Ganz and AEG in Europe and General Electric and Westinghouse Electric in the US. These companies continued to develop AC systems but the technical difference between direct and alternating current systems would follow a much longer technical merger. Due to innovation in the US and Europe, alternating current's economy of scale with very large generating plants linked to loads via long distance transmission was slowly being combined with the ability to link it up with all of the existing systems that needed to be supplied. These included single phase AC systems, poly-phase AC systems, low voltage incandescent lighting, high voltage arc lighting, and existing DC motors in factories and street cars. In what was becoming a *universal system*, these technological differences were temporarily being bridged via the development of rotary converters and motor–generators that would allow the large number of legacy systems to be connected to the AC grid. These stopgaps would slowly be replaced as older systems were retired or upgraded.

Westinghouse alternating current polyphase generators on display at the 1893 World's Fair in Chicago, part of their "Tesla Poly-phase System". Such polyphase innovations revolutionized transmission

The first transmission of three-phase alternating current using high voltage took place in 1891 during the international electricity exhibition in Frankfurt. A 15,000-V transmission line, approximately 175 km long, connected Lauffen on the Neckar and Frankfurt.

Voltages used for electric power transmission increased throughout the 20th century. By 1914, fifty-five transmission systems each operating at more than 70,000-V were in service. The highest voltage then used was 150,000V. By allowing multiple generating plants to be interconnected over a wide area, electricity production cost was reduced. The most efficient available plants could be used to supply the varying loads during the day. Reliability was improved and capital investment cost was reduced, since stand-by generating capacity could be shared over many more customers and a wider geographic area. Remote and low-cost sources of energy, such as hydroelectric power or mine-mouth coal, could be exploited to lower energy production cost.

The rapid industrialization in the 20th century made electrical transmission lines and grids a critical infrastructure item in most industrialized nations. The interconnection of local generation plants and small distribution networks was greatly spurred by the requirements of World War I, with large electrical generating plants built by governments to provide power to munitions factories. Later these generating plants were connected to supply civil loads through long-distance transmission.

BULK POWER TRANSMISSION

A transmission substation decreases the voltage of incoming electricity, allowing it to connect from long distance high voltage transmission, to local lower voltage distribution. It also reroutes power to other transmission lines that serve local markets. This is the PacifiCorp Hale Substation, Orem, Utah, USA

Engineers design transmission networks to transport the energy as efficiently as feasible, while at the same time taking into account economic factors, network safety and redundancy. These networks use components such as power lines, cables, circuit breakers, switches and transformers. The transmission network is usually administered on a regional basis by an entity such as a regional transmission organization or transmission system operator.

Transmission efficiency is greatly improved by devices that increase the voltage (and thereby proportionately reduce the current), in the line conductors, thus allowing power to be transmitted with acceptable losses. The reduced current flowing through the line reduces the heating losses in the conductors. According to Joule's Law, energy losses are directly proportional to the square of the current. Thus, reducing the current by a factor of two will lower the energy lost to conductor resistance by a factor of four for any given size of conductor.

The optimum size of a conductor for a given voltage and current can be estimated by Kelvin's law for conductor size, which states that the size is at its optimum when the annual cost of energy wasted in the resistance is equal to the annual capital charges of providing the conductor. At times of lower interest rates, Kelvin's law indicates that thicker wires are optimal; while, when metals are expensive, thinner conductors are indicated: however, power lines are designed for long-term use, so Kelvin's law has to be used in conjunction with long-term estimates of the price of copper and aluminum as well as interest rates for capital.

The increase in voltage is achieved in AC circuits by using a *step-up transformer*. HVDC systems require relatively costly conversion equipment which may be economically justified for particular projects such as submarine cables and longer distance high capacity point to point transmission. HVDC is necessary for the import and export of energy between grid systems that are not synchronized with each other.

A transmission grid is a network of power stations, transmission lines, and substations. Energy is usually transmitted within a grid with three-phase AC. Single-phase AC is used only for distribution to end users since it is not usable for large polyphase induction motors. In the 19th century, two-phase transmission was used but required either four wires or three wires with unequal currents. Higher order phase systems require more than three wires, but deliver little or no benefit.

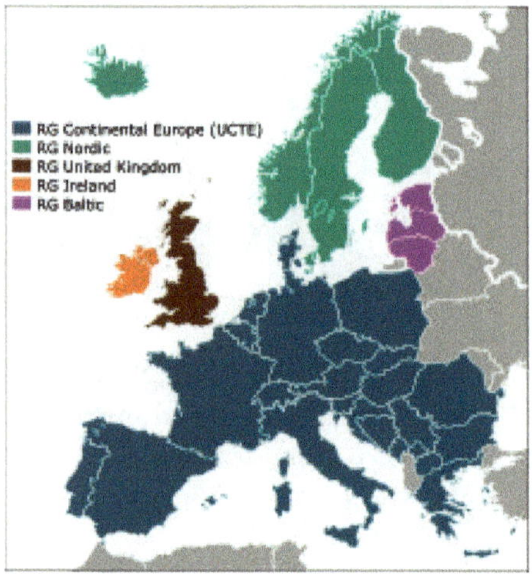

The synchronous grids of the European Union

The price of electric power station capacity is high, and electric demand is variable, so it is often cheaper to import some portion of the needed power than to generate it locally. Because loads are often regionally correlated (hot weather in the Southwest portion of the US might cause many people to use air conditioners), electric power often comes from distant sources. Because of the economic benefits of load sharing between regions, wide area transmission grids now span countries and even continents. The web of interconnections between power producers and consumers should enable power to flow, even if some links are inoperative.

The unvarying (or slowly varying over many hours) portion of the electric demand is known as the *base load* and is generally served by large facilities (which are more efficient due to economies of scale) with fixed costs for fuel and operation. Such facilities are nuclear, coal-fired or hydroelectric, while other energy sources such as concentrated solar thermal and geothermal power have the potential to provide base load power. Renewable energy sources, such as solar photovoltaics, wind, wave, and tidal, are, due to their intermittency, not considered as supplying "base load" but will still add power to the grid. The remaining or 'peak' power demand, is supplied by peaking power plants, which are typically smaller, faster-responding, and higher cost sources, such as combined cycle or combustion turbine plants fueled by natural gas.

Long-distance transmission of electricity (thousands of kilometers) is cheap and efficient, with costs of US$0.005–0.02/kWh (compared to annual averaged large producer costs of US$0.01–0.025/kWh, retail rates upwards of US$0.10/kWh, and multiples of retail for instantaneous suppliers at unpredicted highest demand moments). Thus distant suppliers can be cheaper than local sources (e.g., New York often buys over 1000 MW of electricity from Canada). Multiple **local sources** (even if more expensive and infrequently used) can make the transmission grid more fault tolerant to weather and other disasters that can disconnect distant suppliers.

A high-power electrical transmission tower, 230 kV, double-circuit, also double-bundled

Long-distance transmission allows remote renewable energy resources to be used to displace fossil fuel consumption. Hydro and wind sources cannot be moved closer to populous cities, and solar costs are lowest in remote areas where local power needs are minimal. Connection costs alone can determine whether any particular renewable alternative is economically sensible. Costs can be prohibitive for transmission lines, but various proposals for massive infrastructure investment in high capacity, very long distance super grid transmission networks could be recovered with modest usage fees.

Grid Input

At the power stations, the power is produced at a relatively low voltage between about 2.3 kV and 30 kV, depending on the size of the unit. The generator terminal voltage is then stepped up by the power station transformer to a higher voltage (115 kV to 765 kV AC, varying by the transmission system and by country) for transmission over long distances.

Losses

Transmitting electricity at high voltage reduces the fraction of energy lost to resistance, which varies depending on the specific conductors, the current flowing, and the length of the transmission line. For example, a 100-mile (160 km) 765 kV line carrying 1000 MW of power can have losses of 1.1% to 0.5%. A 345 kV line carrying the same load across the same distance has losses of 4.2%. For a given amount of power, a higher voltage reduces the current and thus the resistive losses in the conductor. For example, raising the voltage by a factor of 10 reduces the current by a corresponding factor of 10 and therefore the I^2R losses by a factor of 100, provided the same sized conductors are used in both cases. Even if the conductor size (cross-sectional area) is reduced 10-fold to match the lower current, the I^2R losses are still reduced 10-fold. Long-distance transmission is typically done with overhead lines at voltages of 115 to 1,200 kV. At extremely high voltages, more than 2,000 kV exists between conductor and ground, corona discharge losses are so large that they can offset the lower resistive losses in the line conductors. Measures to reduce corona losses include conductors having larger diameters; often hollow to save weight, or bundles of two or more conductors.

Factors that affect the resistance, and thus loss, of conductors used in transmission and distribution lines include temperature, spiraling, and the skin effect. The resistance of a conductor increases with its temperature. Temperature changes in electric power lines can have a significant effect on power losses in the line. Spiraling, which refers to the increase in conductor resistance due to the way stranded conductors spiral about the center, also contributes to increases in conductor resistance. The skin effect causes the effective resistance of a conductor to increase at higher alternating current frequencies. Transmission and distribution losses in the USA were estimated at 6.6% in 1997 and 6.5% in 2007. In general, losses are estimated from the discrepancy between power produced (as reported by power plants) and power sold to the end customers; the difference between what is produced and what is consumed constitute transmission and distribution losses, assuming no utility theft occurs.

As of 1980, the longest cost-effective distance for direct-current transmission was determined to be 7,000 km (4,300 mi). For alternating current it was 4,000 km (2,500 mi), though all transmission lines in use today are substantially shorter than this.

In any alternating current transmission line, the inductance and capacitance of the conductors can be significant. Currents that flow solely in 'reaction' to these properties of the circuit, (which together with the resistance define the impedance) constitute reactive power flow, which transmits no 'real' power to the load. These reactive currents, however, are very real and cause extra heating losses in the transmission circuit. The ratio of 'real' power (transmitted to the load) to 'apparent' power (sum of 'real' and 'reactive') is the power factor. As reactive current increases, the reactive power increases and the power factor decreases. For transmission systems with low power factor, losses are higher than for systems with high power factor. Utilities add capacitor banks, reactors and other components (such as phase-shifting transformers; static VAR compensators; and flexible AC transmission systems, FACTS) throughout the system to compensate for the reactive power flow and reduce the losses in power transmission and stabilize system voltages. These measures are collectively called 'reactive support'.

Transposition

Current flowing through transmission lines induces a magnetic field that surrounds the lines of each phase and affects the inductance of the surrounding conductors of other phases. The mutual inductance of the conductors is partially dependent on the physical orientation of the lines with respect to each other. Three-phase power transmission lines are conventionally strung with phases separated on different vertical levels. The mutual inductance seen by a conductor of the phase in the middle of the other two phases will be different than the inductance seen by the conductors on the top or bottom. An imbalanced inductance among the three conductors is problematic because it may result in the middle line carrying a disproportionate amount of the total power transmitted. Similarly, an imbalanced load may occur if one line is consistently closest to the ground and operating at a lower impedance. Because of this phenomenon, conductors must be periodically transposed along the length of the transmission line so that each phase sees equal time in each relative position to balance out the mutual inductance seen by all three phases. To accomplish this, line position is swapped at specially designed transposition towers at regular intervals along the length of the transmission line in various transposition schemes.

Subtransmission

A 115 kV subtransmission line in the Philippines, along with 20 kV distribution lines and a street light, all mounted in a wood subtransmission pole

Subtransmission line tower of wood H-frame construction

Subtransmission is part of an electric power transmission system that runs at relatively lower voltages. It is uneconomical to connect all distribution substations to the high main transmission voltage, because the equipment is larger and more expensive. Typically, only larger substations connect with this high voltage. It is stepped down and sent to smaller substations in towns and neighborhoods. Subtransmission circuits are usually arranged in loops so that a single line failure does not cut off service to a large number of customers for more than a short time. Loops can be "normally closed", where loss of one circuit should result in no interruption, or "normally open" where substations can switch to a backup supply. While subtransmission circuits are usually carried on overhead lines, in urban areas buried cable may be used. The lower-voltage subtransmission lines use less right-of-way and simpler structures; it is much more feasible to put them underground where needed. Higher-voltage lines require more space and are usually above-ground since putting them underground is very expensive.

There is no fixed cutoff between subtransmission and transmission, or subtransmission and distribution. The voltage ranges overlap somewhat. Voltages of 69 kV, 115 kV and 138 kV are often used for subtransmission in North America. As power systems evolved, voltages formerly used for transmission were used for subtransmission, and subtransmission voltages became distribution voltages. Like transmission, subtransmission moves relatively large amounts of power, and like distribution, subtransmission covers an area instead of just point to point.

Transmission Grid Exit

At the substations, transformers reduce the voltage to a lower level for distribution to commercial and residential users. This distribution is accomplished with a combination of sub-transmission (33 kV to 132 kV) and distribution (3.3 to 25 kV). Finally, at the point of use, the energy is transformed to low voltage (varying by country and customer requirements—see Mains electricity by country).

ADVANTAGE OF HIGH-VOLTAGE POWER TRANSMISSION

High-voltage power transmission allows for lesser resistive losses over long distances in the wiring. This efficiency of high voltage transmission allows for the transmission of a larger proportion of the generated power to the substations and in turn to the loads, translating to operational cost savings.

In a very simplified model, assume the electrical grid delivers electricity from a generator (modelled as an ideal voltage source with voltage V, delivering a power P_V) to a single point of consumption, modelled by a pure resistance R, when the wires are long enough to have a significant resistance R_C.

If the resistance are simply in series without any transformer between them, the circuit acts as a voltage divider, because the same current $I = \frac{V}{R+R_C}$ runs through the wire resistance and the powered device. As a consequence, the useful power (used at the point of consumption) is:

$$P_R = V_2 \times I = \frac{V^2}{R+R_C} = \frac{R}{R+R_C}P_V$$

Assume now that a transformer converts high-voltage, low-current electricity transported by the wires into low-voltage, high-current electricity for use at the consumption point. If we suppose it is an ideal transformer with a voltage ratio of a (i.e., the voltage is divided by a and the current is multiplied by a in the secondary branch, compared to the primary branch), then the circuit is again equivalent to a voltage divider, but the transformer-consumption branch has an apparent resistance of a^2R. The useful power is then:

$$P_R = V_2 \times I_2 = \frac{V^2}{a^2R+R_C} = \frac{a^2R}{a^2R+R_C}P_V$$

For $a > 1$ (i.e. conversion of high voltage to low voltage near the consumption point), a larger fraction of the generator's power is transmitted to the consumption point and a lesser fraction is lost to Joule heating.

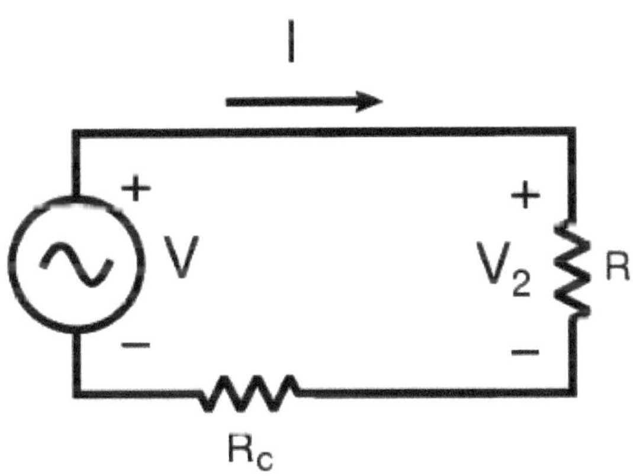

Electrical grid without a transformer.

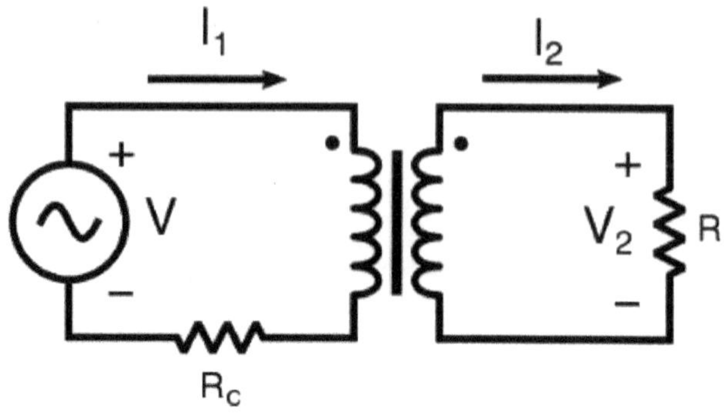

Electrical grid with a transformer.

Modeling: the Transmission Matrix

Oftentimes, we are only interested in the terminal characteristics of the transmission line, which are the voltage and current at the sending and receiving ends. The transmission line itself is then modeled as a "black box" and a 2 by 2 transmission matrix is used to model its behavior, as follows:

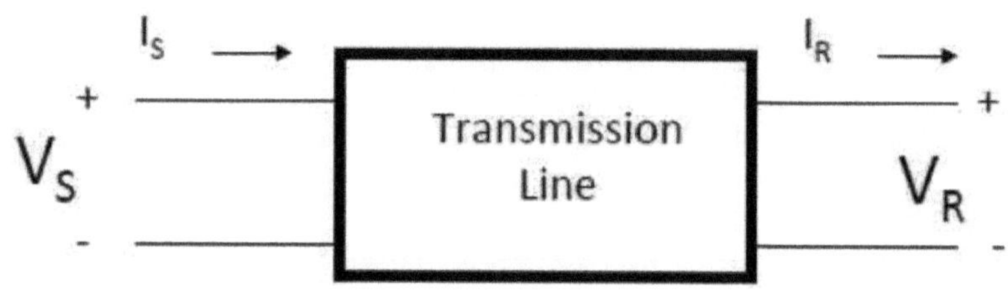

$$\begin{bmatrix} V_S \\ I_S \end{bmatrix} = \begin{bmatrix} A & B \\ C & D \end{bmatrix} \begin{bmatrix} V_R \\ I_R \end{bmatrix}$$

The line is assumed to be a reciprocal, symmetrical network, meaning that the receiving and sending labels can be switched with no consequence. The transmission matrix **T** also has the following properties:

- det(T) = AD - BC = 1
- A = D

The parameters A, B, C, and D differ depending on how the desired model handles the line's resistance (R), inductance (L), capacitance (C), and shunt (parallel) admittance Y. The four main models are the short line approximation, the medium line approximation, the long line approximation (with distributed parameters), and the lossless line. In all models described, a capital letter such as R refers to the total quantity summed over the line and a lowercase letter such as c refers to the per-unit-length quantity.

The **lossless line** approximation, which is the least accurate model, is often used on short lines when the inductance of the line is much greater than its resistance. For this approximation, the voltage and current are identical at the sending and receiving ends.

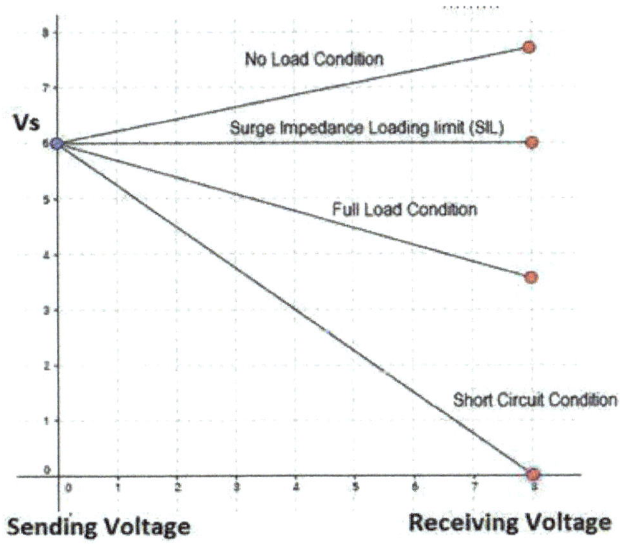

Voltage on sending and receiving ends for lossless line

The characteristic impedance is pure real, which means resistive for that impedance. And it is often called **surge impedance** for a lossless line. When lossless line is terminated by surge impedance, there is no voltage drop. Because though the phase angles of voltage and current are rotated, the magnitudes of voltage and current remain constant along the length of the line. For load > SIL, the voltage will drop from sending end and the line will "consume" VARs. For load < SIL, the voltage will increase from sending end, and the line will "generate" VARs.

The **short line** approximation is normally used for lines less than 50 miles long. For a short line, only a series impedance Z is considered, while C and Y are ignored. The final result is that **A = D = 1 per unit, B = Z ohms**, and **C = 0**.

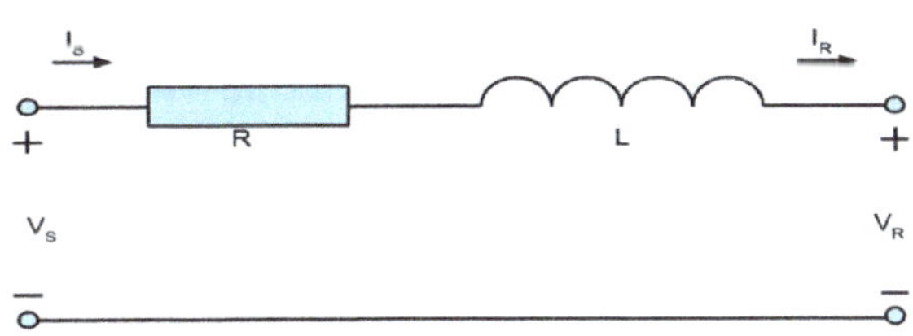

Transmission line for short line approximation

$$\begin{bmatrix} V_S \\ I_S \end{bmatrix} = \begin{bmatrix} 1 & Z \\ 0 & 1 \end{bmatrix} \begin{bmatrix} V_R \\ I_R \end{bmatrix}$$

Transmission matrix for the short line approximation

The **medium line** approximation is used for lines between 50 and 150 miles long. In this model, the series impedance and the shunt admittance are considered, with half of the shunt admittance being placed at each end of the line. This circuit is often referred to as a "nominal pi" circuit because of the shape that is taken on when admittance is placed on both sides of the circuit diagram. The analysis of the medium line brings one to the following result:

$$A = D = 1 + \frac{YZ}{2} \ per\ unit$$

$$B = Z\ \Omega$$

$$C = Y\left(1 + \frac{YZ}{4}\right)\ S$$

Counterintuitive behaviors of medium-length transmission lines:

a) voltage rise at no load or small current

b) receiving end current magnitude can exceed sending end current

The **long line** model is used when a higher degree of accuracy is needed or when the line under consideration is more than 150 miles long. Series resistance and shunt admittance are considered as distributed parameters, meaning each differential length of the line has a corresponding differential resistance and shunt admittance. The following result can be applied at any point along the transmission line, where gamma is defined as the propagation constant.

$$A = D = \cosh(\gamma x)\ per\ unit$$

$$B = Z_c \sinh(\gamma x)\ \Omega$$

$$C = \frac{1}{Z_c} \sinh(\gamma x)\ S$$

To find the voltage and current at the end of the long line, x should be replaced with L (the line length) in all parameters of the transmission matrix.

HIGH-VOLTAGE DIRECT CURRENT

High-voltage direct current (HVDC) is used to transmit large amounts of power over long distances or for interconnections between asynchronous grids. When electrical energy is to be transmitted over very long distances, the power lost in AC transmission becomes appreciable and it is less expensive to use direct current instead of alternating current. For a very long transmission line, these lower losses (and reduced construction cost of a DC line) can offset the additional cost of the required converter stations at each end.

HVDC is also used for submarine cables because AC cannot be supplied over distances of more than about 30 kilometres (19 mi), due to the fact that the cables produce too much reactive power. In these cases special high-voltage cables for DC are used. Submarine HVDC systems are often used to connect the electricity grids of islands, for example, between Great Britain and continental Europe, between Great Britain and Ireland, between Tasmania and the Australian mainland, and between the North and South Islands of New Zealand. Submarine connections up to 600 kilometres (370 mi) in length are presently in use.

HVDC links can be used to control problems in the grid with AC electricity flow. The power transmitted by an AC line increases as the phase angle between source end voltage and destination ends increases, but too large a phase angle will allow the systems at either end of the line to fall out of step. Since the power flow in a DC link is controlled independently of the phases of the AC networks at either end of the link, this phase angle limit does not exist, and a DC link is always able to transfer its full rated power. A DC link therefore stabilizes the AC grid at either end, since power flow and phase angle can then be controlled independently.

As an example, to adjust the flow of AC power on a hypothetical line between Seattle and Boston would require adjustment of the relative phase of the two regional electrical grids. This is an everyday occurrence in AC systems, but one that can become disrupted when AC system components fail and place unexpected loads on the remaining working grid system. With an HVDC line instead, such an interconnection would: (1) Convert AC in Seattle into HVDC; (2) Use HVDC for the 3,000 miles of cross-country transmission; and (3) Convert the HVDC to locally synchronized AC in Boston, (and possibly in other cooperating cities along the transmission route). Such a system could be less prone to failure if parts of it were suddenly shut down. One example of a long DC transmission line is the Pacific DC Intertie located in the Western United States.

Capacity

The amount of power that can be sent over a transmission line is limited. The origins of the limits vary depending on the length of the line. For a short line, the heating of conductors due to line losses sets a thermal limit. If too much current is drawn, conductors may sag too close to the ground, or conductors and equipment may be damaged by overheating. For intermediate-length lines on the order of 100 km (62 mi), the limit is set by the voltage drop in the line. For longer AC lines, system stability sets the limit to the power that can be transferred. Approximately, the power flowing over an AC line is proportional to the cosine of the phase angle of the voltage and current at the receiving and transmitting ends. This angle varies depending on system loading and generation. It is undesirable for the angle to approach 90 degrees, as the power flowing decreases but the resistive losses remain. Very approximately, the allowable product of line length and maximum load is proportional to the square of the system voltage. Series capacitors or phase-shifting transformers are used on long lines to improve stability. High-voltage direct current lines are restricted only by thermal and voltage drop limits, since the phase angle is not material to their operation.

Up to now, it has been almost impossible to foresee the temperature distribution along the cable route, so that the maximum applicable current load was usually set as a compromise between understanding of operation conditions and risk minimization. The availability of industrial distributed temperature sensing (DTS) systems that measure in real time temperatures all along the cable is a first step in monitoring the transmission system capacity. This monitoring solution is based on using passive optical fibers as temperature sensors, either integrated directly inside a high voltage cable or mounted externally on the cable insulation. A solution for overhead lines is also available. In this case the optical fiber is integrated into the core of a phase wire of overhead transmission lines (OPPC). The integrated Dynamic Cable Rating (DCR) or also called Real Time Thermal Rating (RTTR) solution enables not only to continuously monitor the temperature of a high voltage cable circuit in real time, but to safely utilize the existing network capacity to its maximum. Furthermore, it provides the ability to the operator to predict the behavior of the transmission system upon major changes made to its initial operating conditions.

Control

To ensure safe and predictable operation, the components of the transmission system are controlled with generators, switches, circuit breakers and loads. The voltage, power, frequency, load factor, and reliability capabilities of the transmission system are designed to provide cost effective performance for the customers.

Load Balancing

The transmission system provides for base load and peak load capability, with safety and fault tolerance margins. The peak load times vary by region largely due to the industry mix. In very hot and very cold climates home air conditioning and heating loads have an effect on the overall load. They are typically highest in the late afternoon in the hottest part of the year and in mid-mornings and mid-evenings in the coldest part of the year. This makes the power requirements vary by the season and the time of day. Distribution system designs always take the base load and the peak load into consideration.

The transmission system usually does not have a large buffering capability to match the loads with the generation. Thus generation has to be kept matched to the load, to prevent overloading failures of the generation equipment.

Multiple sources and loads can be connected to the transmission system and they must be controlled to provide orderly transfer of power. In centralized power generation, only local control of generation is necessary, and it involves synchronization of the generation units, to prevent large transients and overload conditions.

In distributed power generation the generators are geographically distributed and the process to bring them online and offline must be carefully controlled. The load control signals can either be sent on separate lines or on the power lines themselves. Voltage and frequency can be used as signalling mechanisms to balance the loads.

In voltage signaling, the variation of voltage is used to increase generation. The power added by any system increases as the line voltage decreases. This arrangement is stable in principle. Voltage-based regulation is complex to use in mesh networks, since the individual components and setpoints would need to be reconfigured every time a new generator is added to the mesh.

In frequency signaling, the generating units match the frequency of the power transmission system. In droop speed control, if the frequency decreases, the power is increased. (The drop in line frequency is an indication that the increased load is causing the generators to slow down.)

Wind turbines, vehicle-to-grid and other distributed storage and generation systems can be connected to the power grid, and interact with it to improve system operation.

Failure Protection

Under excess load conditions, the system can be designed to fail gracefully rather than all at once. Brownouts occur when the supply power drops below the demand. Blackouts occur when the supply fails completely.

Rolling blackouts (also called load shedding) are intentionally engineered electrical power outages, used to distribute insufficient power when the demand for electricity exceeds the supply.

Communications

Operators of long transmission lines require reliable communications for control of the power grid and, often, associated generation and distribution facilities. Fault-sensing protective relays at each end of the line must communicate to monitor the flow of power into and out of the protected line section so that faulted conductors or equipment can be quickly de-energized and the balance of the system restored. Protection of the transmission line from short circuits and other faults is usually so critical that common carrier telecommunications are insufficiently reliable, and in remote areas a common carrier may not be available. Communication systems associated with a transmission project may use:

- Microwaves
- Power line communication
- Optical fibers

Rarely, and for short distances, a utility will use pilot-wires strung along the transmission line path. Leased circuits from common carriers are not preferred since availability is not under control of the electric power transmission organization.

Transmission lines can also be used to carry data: this is called power-line carrier, or PLC. PLC signals can be easily received with a radio for the long wave range.

Optical fibers can be included in the stranded conductors of a transmission line, in the overhead shield wires. These cables are known as optical ground wire (*OPGW*). Sometimes a standalone cable is used, all-dielectric self-supporting (*ADSS*) cable, attached to the transmission line cross arms.

Some jurisdictions, such as Minnesota, prohibit energy transmission companies from selling surplus communication bandwidth or acting as a telecommunications common carrier. Where the regulatory structure permits, the utility can sell capacity in extra dark fibers to a common carrier, providing another revenue stream.

Electricity Market Reform

Some regulators regard electric transmission to be a natural monopoly and there are moves in many countries to separately regulate transmission (see electricity market).

Spain was the first country to establish a regional transmission organization. In that country, transmission operations and market operations are controlled by separate companies. The transmission system operator is Red Eléctrica de España (REE) and the wholesale electricity market operator is Operador del Mercado Ibérico de Energía – Polo Español, S.A. (OMEL) . Spain's transmission system is interconnected with those of France, Portugal, and Morocco.

In the United States and parts of Canada, electrical transmission companies operate independently of generation and distribution companies.

Cost of Electric Power Transmission

The cost of high voltage electricity transmission (as opposed to the costs of electric power distribution) is comparatively low, compared to all other costs arising in a consumer's electricity bill. In the UK, transmission costs are about 0.2p/kWh compared to a delivered domestic price of around 10p/kWh.

Research evaluates the level of capital expenditure in the electric power T&D equipment market will be worth $128.9bn in 2011.

Merchant Transmission

Merchant transmission is an arrangement where a third party constructs and operates electric transmission lines through the franchise area of an unrelated utility.

Operating merchant transmission projects in the United States include the Cross Sound Cable from Shoreham, New York to New Haven, Connecticut, Neptune RTS Transmission Line from Sayreville, N.J., to Newbridge, N.Y, and Path 15 in California. Additional projects are in development or have been proposed throughout the United States, including the Lake Erie Connector, an underwater transmission line proposed by ITC Holdings Corp., connecting Ontario to load serving entities in the PJM Interconnection region.

There is only one unregulated or market interconnector in Australia: Basslink between Tasmania and Victoria. Two DC links originally implemented as market interconnectors, Directlink and Murraylink, have been converted to regulated interconnectors. NEMMCO

A major barrier to wider adoption of merchant transmission is the difficulty in identifying who benefits from the facility so that the beneficiaries will pay the toll. Also, it is difficult for a merchant transmission line to compete when the alternative transmission lines are subsidized by other utility businesses.

Health Concerns

Some large studies, including a large study in the United States, have failed to find any link between living near power lines and developing any sickness or diseases, such as cancer. A 1997 study found that it did not matter how close one was to a power line or a sub-station, there was no increased risk of cancer or illness.

The mainstream scientific evidence suggests that low-power, low-frequency, electromagnetic radiation associated with household currents and high transmission power lines does not constitute a short or long term health hazard. Some studies, however, have found statistical correlations between various diseases and living or working near power lines. No adverse health effects have been substantiated for people not living close to powerlines.

There are established biological effects for acute *high* level exposure to magnetic fields well above 100 μT (1 G). In a residential setting, there is "limited evidence of carcinogenicity in humans and less than sufficient evidence for carcinogenicity in experimental animals", in particular, childhood leukemia, *associated with* average exposure to residential power-frequency magnetic field above 0.3 μT (3 mG) to 0.4 μT (4 mG). These levels exceed average residential power-frequency magnetic fields in homes, which are about 0.07 μT (0.7 mG) in Europe and 0.11 μT (1.1 mG) in North America.

The Earth's natural geomagnetic field strength varies over the surface of the planet between 0.035 mT - 0.07 mT (35 μT - 70 μT or 0.35 G - 0.7 G) while the International Standard for the continuous exposure limit is set at 40 mT (40,000 μT or 400 G) for the general public.

Tree Growth Regulator and Herbicide Control Methods may be used in transmission line right of ways which may have health effects.

REFERENCES

1. "A Primer on Electric Utilities, Deregulation, and Restructuring of U.S. Electricity Markets" (pdf). United States Department of Energy Federal Energy Management Program (FEMP). May 2002. Retrieved December 27, 2008.
2. Hans Dieter Betz, Ulrich Schumann, Pierre Laroche (2009). Lightning: Principles, Instruments and Applications.Springer, pp. 202–203. ISBN 978-1-4020-9078-3. Retrieved on May 13, 2009.
3. Thomas P. Hughes (1993). Networks of Power: Electrification in Western Society, 1880–1930. Baltimore: Johns Hopkins University Press. pp. 119–122. ISBN 0-8018-4614-5.
4. Guarnieri, M. (2013). "The Beginning of Electric Energy Transmission: Part One". IEEE Industrial Electronics Magazine. 7 (1): 57–60. doi:10.1109/MIE.2012.2236484.
5. National Council on Electricity Policy. "Electricity Transmission: A primer" (pdf).
6. Guarnieri, M. (2013). "The Beginning of Electric Energy Transmission: Part Two". IEEE Industrial Electronics Magazine. 7 (2): 52–59. doi:10.1109/MIE.2013.2256297.
7. Great Barrington 1886 - Inspiring an industry toward AC power
8. ethw.org - William Stanley, Jr.
9. Arnold Heertje, Mark Perlman Evolving Technology and Market Structure: Studies in Schumpeterian Economics, page 138
10. Parke Hughes, Thomas (1993). Networks of Power: Electrification in Western Society, 1880-1930. JHU Press. pp. 120–121.
11. Garud, Raghu; Kumaraswamy, Arun; Langlois, Richard (2009). Managing in the Modular Age: Architectures, Networks, and Organizations. John Wiley & Sons. p. 249.
12. Kiessling F, Nefzger P, Nolasco JF, Kaintzyk U (2003) Overhead power lines. Springer, Berlin, Heidelberg, New York, p. 5
13. Bureau of Census data reprinted in Hughes, pp. 282–283
14. Hughes, pp. 293–295
15. Paris, L.; Zini, G.; Valtorta, M.; Manzoni, G.; Invernizzi, A.; De Franco, N.; Vian, A. (1984). "Present Limits of Very Long Distance Transmission Systems" (pdf). CIGRE International Conference on Large High Voltage Electric Systems,

1984 Session, 29th August-6th September. Global Energy Network Institute. Retrieved March 29, 2011. 4.98 MB

16. "NYISO Zone Maps". New York Independent System Operator. Retrieved January 10, 2014.
17. American Electric Power, Transmission Facts, page 4: http://www.aep.com/about/transmission/docs/transmission-facts.pdf
18. California Public Utilities Commission Corona and induced currents
19. "Where can I find data on electricity transmission and distribution losses?". Frequently Asked Questions – Electricity. U.S. Energy Information Administration. November 19, 2009. Retrieved March 29, 2011.
20. Donald G. Fink and H. Wayne Beaty, Standard Handbook for Electrical Engineers (15th Edition) McGraw-Hill, 2007 ISBN 978-0-07-144146-9 section 18.5
21. Guarnieri, M. (2013). "The Alternating Evolution of DC Power Transmission". IEEE Industrial Electronics Magazine. 7 (3): 60–63. doi:10.1109/MIE.2013.2272238.
22. Raghuvir Srinivasan (August 15, 2004). "Power transmission business is a natural monopoly". The Hindu Business Line. The Hindu. Retrieved January 31, 2008.
23. Lynne Kiesling (August 18, 2003). "Rethink the Natural Monopoly Justification of Electricity Regulation". Reason Foundation. Retrieved January 31, 2008.
24. What is the cost per kWh of bulk transmission / National Grid in the UK (note this excludes distribution costs)
25. The Electric Power Transmission & Distribution (T&D) Equipment Market 2011–2021
26. How ITC Holdings plans to connect PJM demand with Ontario's rich renewables, Utility Dive, Dec. 8, 2014, http://www.utilitydive.com/news/how-itc-holdings-plans-to-connect-pjm-demand-with-ontarios-rich-renewables/341524/
27. Fiona Woolf (February 2003). Global Transmission Expansion. Pennwell Books. pp. 226, 247. ISBN 0-87814-862-0.
28. Power Lines and Cancer, The Health Report / ABC Science - Broadcast on June 7, 1997 (Australian Broadcasting Corporation)
29. Electromagnetic fields and public health, World Health Organization
30. "Electromagnetic fields and public health". Fact sheet No. 322. World Health Organization. June 2007. Retrieved January 23, 2008.
31. "Electric and Magnetic Fields Associated with the Use of Power" (PDF). National Institute of Environmental Health Sciences. June 2002. Retrieved January 29, 2008.
32. Transmission Vegetation Management NERC Standard FAC-003-2 Technical Reference Page 14/50. http://www.nerc.com/docs/standards/sar/FAC-003-2_White_Paper_2009Sept9.pdf
33. "Order No. 888". United States of America Federal Energy Regulatory Commission.

34. Order No. 888, FERC. https://www.ferc.gov/legal/maj-ord-reg/land-docs/order888.asp. Missing or empty |title=(help)
35. Energy Policy Act of 2005 Fact Sheet (PDF). FERC Washington, D.C. August 8, 2006.
36. National Council on Electricity Policy. "Electricity Transmission: A primer" (pdf): 32 (41 in pdf).
37. Wald, Matthew (August 27, 2008). "Wind Energy Bumps Into Power Grid's Limits". New York Times: A1. Retrieved December 12, 2008.
38. Jacob Oestergaard; et al. (2001). "Energy losses of superconducting power transmission cables in the grid". IEEE Transactions on Applied Superconductivity. 11: 2375. doi:10.1109/77.920339.
39. 600m superconducting electricity line laid in New York
40. Superconducting cables will be used to supply electricity to consumers
41. Superconductivity's First Century
42. Albany HTS Cable Project
43. High-Temperature Superconductors
44. High-Temperature Superconductor Technology Stepped Up
45. Operation of longest superconducting cable worldwide started
46. BBC: Spies 'infiltrate US power grid'
47. CNN: Video
48. Reuters: US concerned power grid vulnerable to cyber-attack
49. "Energy Systems, Environment and Development". Advanced Technology Assessment Systems. Global Energy Network Institute (6). Autumn 1991. Retrieved December 27, 2008.
50. "India Steps It Up". Transmission & Distribution World. January 2013.

CHAPTER 9

Instrumentation

Chapter Contents

Instrumentation is a specialised branch of Electrical and Electronics Engineering, primarily focussing on the principles and operations of measuring instruments used in the design and configuration of automated systems. Instrumentation technology helps create, construct and maintain measuring and control devices and systems found in manufacturing plants and research institutions.

INTRODUCTION

Instrumentation is a collective term for measuring instruments used for indicating, measuring and recording physical quantities.

The term instrumentation may refer to something as simple as direct reading thermometers or, when using many sensors, may become part of a complex Industrial control system in such as manufacturing industry, vehicles and transportation. Instrumentation can be found in the household as well; a smoke detector or a heating thermostat are examples.

PRE-INDUSTRIAL

Elements of industrial instrumentation have long histories. Scales for comparing weights and simple pointers to indicate position are ancient technologies. Some of the earliest measurements were of time. One of the oldest water clocks was found in the tomb of the Egyptian pharaoh Amenhotep I, buried around 1500 BCE.Improvements were incorporated in the clocks. By 270 BCE they had the rudiments of an automatic control system device.

In 1663 Christopher Wren presented the Royal Society with a design for a "weather clock". A drawing shows meteorological sensors moving pens over paper driven by clockwork. Such devices did not become standard in meteorology for two centuries.The concept has remained virtually unchanged as evidenced by pneumatic chart recorders, where a pressurized bellows displaces a pen. Integrating sensors, displays, recorders and controls was uncommon until the industrial revolution, limited by both need and practicality.

A local instrumentation panel on a steam turbine.

Early Industrial

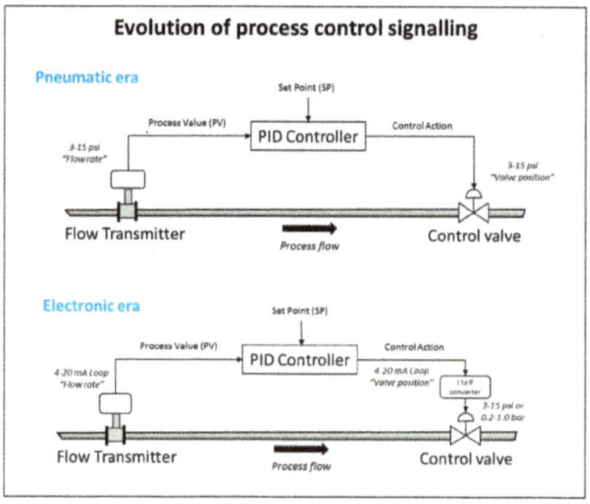

The evolution of analogue control loop signalling from the pneumatic era to the electronic era.

Electronics enabled wiring to replace pipes. A transmitter is a device that produces an output signal, often in the form of a 4–20 mA electrical current signal, although many other options using voltage, frequency, pressure, or ethernet are possible. The transistor was commercialized by the mid-1950s.

Instruments attached to a control system provided signals used to operate solenoids, valves, regulators, circuit breakers, relays and other devices. Such devices could control a desired output variable, and provide either remote or automated control capabilities.

Each instrument company introduced their own standard instrumentation signal, causing confusion until the 4-20 mA range was used as the standard electronic instrument signal for transmitters and valves. This signal was eventually standardized as ANSI/ISA S50, "Compatibility of Analog Signals for Electronic Industrial Process Instruments», in the 1970s. The transformation of instrumentation from mechanical pneumatic transmitters, controllers, and valves to electronic instruments reduced maintenance costs as electronic instruments were more dependable than mechanical instruments. This also increased efficiency and production due to their increase in accuracy. Pneumatics enjoyed some advantages, being favored in corrosive and explosive atmospheres.

Automatic Process Control

In the early years of process control, process indicators and control elements such as valves were monitored by an operator that walked around the unit adjusting the valves to obtain the desired temperatures, pressures, and flows. As technology evolved pneumatic controllers were invented and mounted in the field that monitored the process and controlled the valves. This reduced the amount of time process operators were needed to monitor the process. Later years the actual controllers were moved to a central room and signals were sent into the control room to monitor the process and outputs signals were sent to the final control element such as a valve to adjust the process as needed. These controllers and indicators were mounted on a wall called a control board. The operators stood in front of this board walking back and forth monitoring the process indicators. This again reduced the number and amount of time process operators were needed to walk around the units. The most standard pneumatic signal level used during these years was 3-15 psig.

Large Integrated Computer-Based Systems

Pneumatic "Three term" pneumatic PID controller, widely used before electronics became reliable and cheaper

A pre-DCS/SCADA era central control room. Whilst the controls are centralised in one place, they are still discrete and not integrated into one system.

A DCS control room where plant information and controls are displayed on computer graphics screens. The operators are seated and can view and control any part of the process from their screens, whilst retaining a plant overview.

Process control of large industrial plants has evolved through many stages. Initially, control would be from panels local to the process plant. However this required a large manpower resource to attend to these dispersed panels, and there was no overall view of the process. The next logical development was the transmission of all plant measurements to a permanently-manned central control room. Effectively this was the centralisation of all the localised panels, with the advantages of lower manning levels and easier overview of the process. Often the controllers were behind the control room panels, and all automatic and manual control outputs were transmitted back to plant.

However, whilst providing a central control focus, this arrangement was inflexible as each control loop had its own controller hardware, and continual operator movement within the control room was required to view different parts of the process. With coming of electronic processors and graphic displays it became possible to replace these discrete controllers with computer-based algorithms, hosted on a network of input/output racks with their own control processors. These could be distributed around plant, and communicate with the graphic display in the control room or rooms. The distributed control concept was born.

The introduction of DCSs and SCADA allowed easy interconnection and re-configuration of plant controls such as cascaded loops and interlocks, and easy interfacing with other production computer systems. It enabled sophisticated alarm handling, introduced automatic event logging, removed the need for physical records such as chart recorders, allowed the control racks to be networked and thereby located locally to plant to reduce cabling runs, and provided high level overviews of plant status and production levels.

APPLICATIONS

In some cases the sensor is a very minor element of the mechanism. Digital cameras and wristwatches might technically meet the loose definition of instrumentation because they record and/or display sensed information. Under most circumstances neither would be called instrumentation, but when used to measure the elapsed time of a race and to document the winner at the finish line, both would be called instrumentation.

Household

A very simple example of an instrumentation system is a mechanical thermostat, used to control a household furnace and thus to control room temperature. A typical unit senses temperature with a bi-metallic strip. It displays temperature by a needle on the free end of the strip. It activates the furnace by a mercury switch. As the switch is rotated by the strip, the mercury makes physical (and thus electrical) contact between electrodes.

Another example of an instrumentation system is a home security system. Such a system consists of sensors (motion detection, switches to detect door openings), simple algorithms to detect intrusion, local control (arm/disarm) and remote monitoring of the system so that the police can be summoned. Communication is an inherent part of the design.

Kitchen appliances use sensors for control.

- A refrigerator maintains a constant temperature by measuring the internal temperature.
- A microwave oven sometimes cooks via a heat-sense-heat-sense cycle until sensing done.
- An automatic ice machine makes ice until a limit switch is thrown.
- Pop-up bread toasters can operate by time or by heat measurements.
- Some ovens use a temperature probe to cook until a target internal food temperature is reached.

- A common toilet refills the water tank until a float closes the valve. The float is acting as a water level sensor.

Automotive

Modern automobiles have complex instrumentation. In addition to displays of engine rotational speed and vehicle linear speed, there are also displays of battery voltage and current, fluid levels, fluid temperatures, distance traveled and feedbacks of various controls (turn signals, parking brake, headlights, transmission position). Cautions may be displayed for special problems (fuel low, check engine, tire pressure low, door ajar, seat belt unfastened). Problems are recorded so they can be reported to diagnostic equipment. Navigation systems can provide voice commands to reach a destination. Automotive instrumentation must be cheap and reliable over long periods in harsh environments. There may be independent airbag systems which contain sensors, logic and actuators. Anti-skid braking systems use sensors to control the brakes, while cruise control affects throttle position. A wide variety of services can be provided via communication links as the OnStar system. Autonomous cars (with exotic instrumentation) have been demonstrated.

Aircraft

Early aircraft had a few sensors. "Steam gauges" converted air pressures into needle deflections that could be interpreted as altitude and airspeed. A magnetic compass provided a sense of direction. The displays to the pilot were as critical as the measurements.

A modern aircraft has a far more sophisticated suite of sensors and displays, which are embedded into avionics systems. The aircraft may contain inertial navigation systems, global positioning systems, weather radar, autopilots, and aircraft stabilization systems. Redundant sensors are used for reliability. A subset of the information may be transferred to a crash recorder to aid mishap investigations. Modern pilot displays now include computer displays including head-up displays.

Air traffic control radar is distributed instrumentation system. The ground portion transmits an electromagnetic pulse and receives an echo (at least). Aircraft carry transponders that transmit codes on reception of the pulse. The system displays aircraft map location, an identifier and optionally altitude. The map location is based on sensed antenna direction and sensed time delay. The other information is embedded in the transponder transmission.

Laboratory Instrumentation

Among the possible uses of the term is a collection of laboratory test equipment controlled by a computer through an IEEE-488 bus (also known as GPIB for General Purpose Instrument Bus or HPIB for Hewlitt Packard Instrument Bus). Laboratory equipment is available to measure many electrical and chemical quantities. Such a collection of equipment might be used to automate the testing of drinking water for pollutants.

Measurement Parameters

Instrumentation is used to measure many parameters (physical values). These parameters include:

- Pressure, either differential or static
- Flow
- Temperature
- Levels of liquids, etc.
- Density
- Viscosity
- ionizing radiation
- Frequency
- Current
- Voltage
- Inductance
- Capacitance
- Resistivity
- Chemical composition
- Chemical properties
- Vibration
- Weight

Instrumentation Engineering

Control valve.

Instrumentation engineering is the engineering specialization focused on the principle and operation of measuring instruments that are used in design and configuration of automated systems in electrical, pneumatic domains etc. They typically work for industries with automated processes, such as chemical or manufacturing plants, with the goal of improving system productivity, reliability, safety, optimization, and stability. To control the parameters in a process or in a particular system, devices such as microprocessors, microcontrollers or PLCs are used, but their ultimate aim is to control the parameters of a system.

Instrumentation engineering is loosely defined because the required tasks are very domain dependent. An expert in the biomedical instrumentation of laboratory rats has very different concerns than the expert in rocket instrumentation. Common concerns of both are the selection of appropriate sensors based on size, weight, cost, reliability, accuracy, longevity, environmental robustness and frequency response. Some sensors are literally fired in artillery shells. Others sense thermonuclear explosions until destroyed. Invariably sensor data must be recorded, transmitted or displayed. Recording rates and capacities vary enormously. Transmission can be trivial or can be clandestine, encrypted and low-power in the presence of jamming. Displays can be trivially simple or can require consultation with human factors experts. Control system design varies from trivial to a separate specialty.

Instrumentation engineers are responsible for integrating the sensors with the recorders, transmitters, displays or control systems, and producing the Piping and instrumentation diagram for the process. They may design or specify installation, wiring and signal conditioning. They may be responsible for calibration, testing and maintenance of the system.

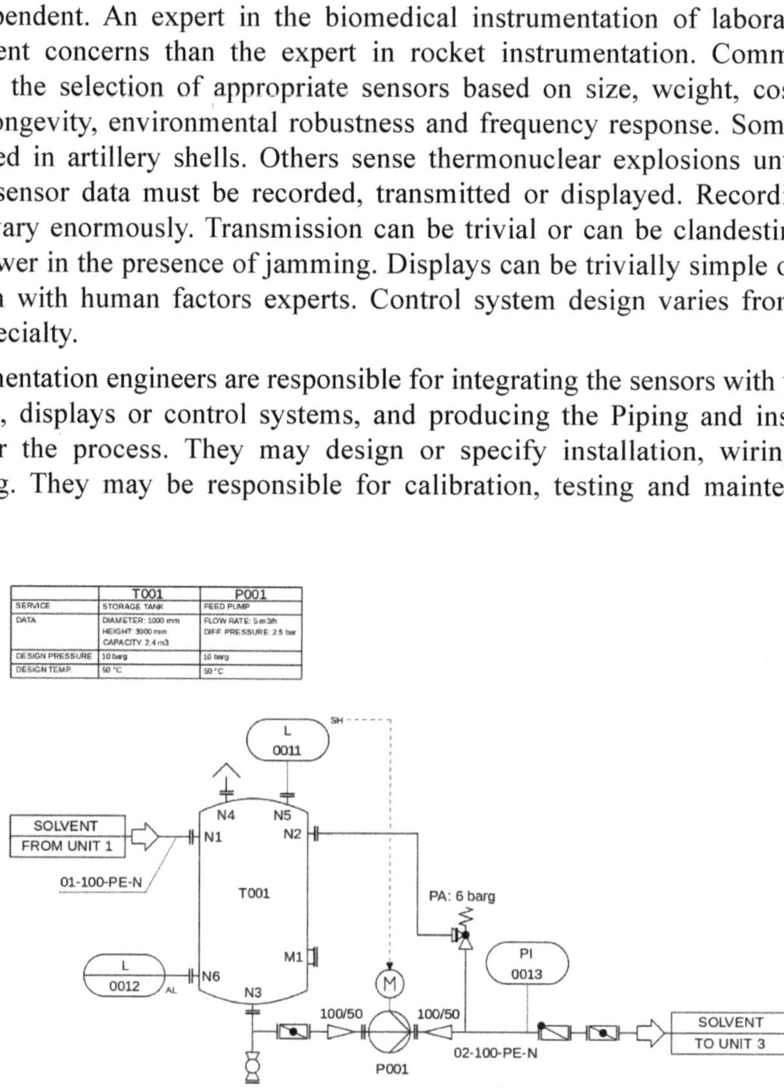

The instrumentation part of a Piping and instrumentation diagram will be developed by an instrumentation engineer.

In a research environment it is common for subject matter experts to have substantial instrumentation system expertise. An astronomer knows the structure of the universe and a great deal about telescopes - optics, pointing and cameras (or other sensing elements). That often includes the hard-won knowledge of the operational procedures that provide the best results. For example, an astronomer is often knowledgeable of techniques to minimize temperature gradients that cause air turbulence within the telescope.

Instrumentation technologists, technicians and mechanics specialize in troubleshooting, repairing and maintaining instruments and instrumentation systems.

Typical Industrial Transmitter Signal Types

Current Loop (4-20mA) - Electrical

HART - Data signalling often overlaid on a current loop.

Foundation Fieldbus - Data signalling

Profibus - Data signalling

Impact of Modern Development

Ralph Müller (1940) stated "That the history of physical science is largely the history of instruments and their intelligent use is well known. The broad generalizations and theories which have arisen from time to time have stood or fallen on the basis of accurate measurement, and in several instances new instruments have had to be devised for the purpose. There is little evidence to show that the mind of modern man is superior to that of the ancients. His tools are incomparably better."

Davis Baird has argued that the major change associated with Floris Cohen's identification of a "fourth big scientific revolution" after World War II is the development of scientific instrumentation, not only in chemistry but across the sciences. In chemistry, the introduction of new instrumentation in the 1940s was "nothing less than a scientific and technological revolution" in which classical wet-and-dry methods of structural organic chemistry were discarded, and new areas of research opened up.

As early as 1954, W A Wildhack discussed both the productive and destructive potential inherent in process control. The ability to make precise, verifiable and reproducible measurements of the natural world, at levels that were not previously observable, using scientific instrumentation, has "provided a different texture of the world". This instrumentation revolution fundamentally changes human abilities to monitor and respond, as is illustrated in the examples of DDT monitoring and the use of UV spectrophotometry and gas chromatography to monitor water pollutants.

REFERENCES

1. "Early Clocks". Retrieved 1 March 2012.
2. "Building automation history page". Retrieved 1 March 2012.

3. Multhauf, Robert P. (1961), The Introduction of Self-Registering Meteorological Instruments, Washington, D.C.: Smithsonian Institution, pp. 95–116 United States National Museum, Bulletin 228. Contributions from The Museum of History and Technology: Paper 23. Available from Project Gutenberg.
4. Archived May 18, 2015, at the Wayback Machine.
5. Anderson, Norman A. (1998). Instrumentation for Process Measurement and Control (3 ed.). CRC Press. pp. 254–255. ISBN 0-8493-9871-1.
6. Anderson, Norman A. (1998). Instrumentation for Process Measurement and Control (3 ed.). CRC Press. pp. 8–10. ISBN 0-8493-9871-1.
7. Aircraft Instrumentation - Leroy R. Grumman Cadet Squadron
8. Katz, Eric; Light, Andrew; Thompson, William (2002). Controlling technology : contemporary issues (2nd ed.). Amherst, NY: Prometheus Books. ISBN 978-1573929837. Retrieved 9 March 2016.
9. Baird, D. (1993). "Analytical chemistry and the 'big' scientific instrumentation revolution". Annals of Science. 50: 267–290. doi:10.1080/00033799300200221. Download the pdf to read the full article.
10. Baird, D. (2002). "Analytical chemistry and the 'big' scientific instrumentation revolution". In Morris, Peter J. T. From classical to modern chemistry : the instrumental revolution ; from a conference on the history of chemical instrumentation: "From the Test-tube to the Autoanalyzer: the Development of Chemical Instrumentation in the Twentieth Century", London, in August 2000. Cambridge: Royal Society of Chemistry in assoc. with the Science Museum. pp. 29–56. ISBN 9780854044795.
11. Reinhardt, Carsten, ed. (2001). Chemical sciences in twentieth century (1st ed.). Weinheim: Wiley-VCH. ISBN 978-3527302710.
12. Wildhack, W. A. (22 October 1954). "Instrumentation--Revolution in Industry, Science, and Warfare". Science. 120 (3121): 15A–15A. doi:10.1126/science.120.3121.15A. Retrieved 9 March 2016.
13. Hentschel, Klaus (2003). "The Instrumental Revolution in Chemistry (Review Essay)". Foundations of Chemistry. 5 (2): 179–183. doi:10.1023/A:1023691917565. Retrieved 8 March 2016.

Index

M

P

R

S

T

U

V